Evolutionary Biology

VOLUME 31

Evolutionary Biology

VOLUME 31

Edited by

MAX K. HECHT

*Queens College of the
City University of New York
Flushing, New York*

ROSS J. MACINTYRE

*Cornell University
Ithaca, New York*

and

MICHAEL T. CLEGG

*University of California, Riverside
Riverside, California*

KLUWER ACADEMIC / PLENUM PUBLISHERS
NEW YORK, BOSTON, DORDRECHT, LONDON, MOSCOW

The Library of Congress catalogued the first volume of this title as follows:

Evolutionary biology, v. 1–   1967–

New York, Appleton-Century-Crofts.
v. illus., 24 cm annual.
Editors: 1967–   T. Dobzhansky and others.
1. Evolution—Period.   2. Biology—Period.   I. Dobzhanksy, Theodosius
Grigorievich, 1900–

QH366.A1E9   575'.005   67-11961

ISBN: 0-306-46178-1

© 2000 Kluwer Academic / Plenum Publishers
233 Spring Street, New York, N.Y. 10013

10 9 8 7 6 5 4 3 2

A C.I.P. record for this book is available from the Library of Congress

Printed in the United States of America

Contributors

Elysse M. Craddock • *Division of Natural Sciences, Purchase College, State University of New York, Purchase, New York 10577-1400*

Jerzy Dzik • *Institute of Paleobiology PAN, 00-818 Warsaw, Poland*

Brian R. Morton • *Department of Biological Sciences, Barnard College, Columbia University, New York, New York 10027*

Kurt Schwenk • *Department of Ecology and Evolutionary Biology, University of Connecticut, Storrs, Connecticut 06269-3043*

Günter P. Wagner • *Department of Ecology and Evolutionary Biology and Center for Computational Ecology, Yale University, New Haven, Connecticut 06520-8106*

Preface

This volume is the thirty-second in this series, which includes thirty-one numbered volumes and one unnumbered supplement. The editors continue to focus on critical reviews, commentaries, original papers, and controversies in evolutionary biology. The topics of the review range from developmental biology to paleobiology.

Recent volumes have included a broad spectrum of chapters on such subjects as molecular phylogenetics, homology and embryonic development, and paleobiological and developmental aspects of limb development.

The editors continue to solicit manuscripts in all areas of evolutionary biology. Manuscripts should be sent to any one of the following: Max K. Hecht, Department of Biology, Queens College of CUNY, Flushing, New York 11367; Ross J. MacIntyre, Department of Genetics and Development, Cornell University, Ithaca, New York 14853; or Michael T. Clegg, Department of Botany and Plant Sciences, University of California, Riverside, California 92521.

Contents

3. The Origin of the Mineral Skeleton in Chordates

Jerzy Dzik

4. Evolutionarily Stable Configurations: Functional Integration and the Evolution of Phenotypic Stability

Günter P. Wagner and Kurt Schwenk

1

Speciation Processes in the Adaptive Radiation of Hawaiian Plants and Animals

ELYSSE M. CRADDOCK

INTRODUCTION

Hawaii, the most isolated archipelago on earth, is an evolutionist's paradise. Despite its comparative geological recency and extreme isolation, this chain of volcanic islands in the northern Pacific hosts an amazingly rich and diverse biota—the outcome of rapid, explosive speciation and adaptive radiation in many groups of plants and animals. Most of the flora and fauna are endemic to Hawaii, with endemism rates as high as 90 to 99% for terrestrial forms (Carlquist, 1970); in fact, many species are restricted to individual islands or volcanoes. Thus it can be inferred that these endemic forms speciated *in situ*, following successful colonization by a series of founder individuals.

The original founders arrived by rare chance over a time period of many millions of years, via various means of long distance dispersal (Carlquist, 1970, 1982) from Indo-West Pacific sources and to a lesser extent from North America. Descendants of the founding populations established on older islands in the chain subsequently provided founder individuals that colonized the younger islands as they became habitable. Successive island-hopping events dispersing life throughout the Hawaiian Archipelago have

ELYSSE M. CRADDOCK • Division of Natural Sciences, Purchase College, State University of New York, Purchase, New York 10577-1400.

Evolutionary Biology, Volume 31, edited by Max K. Hecht *et al.* Kluwer Academic / Plenum Publishers, New York, 2000.

been traced using characters ranging from morphology (Wagner *et al.*, 1995; Shaw, 1995) to chromosomes (Carson, 1970, 1983), chloroplast DNA restriction site variation (Baldwin *et al.*, 1990; Givnish *et al.*, 1995; P. S. Soltis *et al.*, 1996), sequences of mitochondrial (DeSalle, 1995; Shaw, 1996b) and nuclear DNA (Kambysellis *et al.*, 1995), and even behavior (Kaneshiro, 1983). Significantly, movement of these interisland founders appears to have been largely unidirectional, from older islands to the northwest to younger islands to the southeast (Funk and Wagner, 1995). The unique feature of these insular founder events is that so many have triggered the origin of new species, leading to rampant speciation in Hawaiian plants and animals. It is incontrovertible that speciation in Hawaii is somehow tied to founder events. The questions are how it is tied to these events and what biologists can learn about the speciation process from analyzing extant organisms in this unparalleled natural laboratory and crucible for evolutionary change.

The divergence of life forms on oceanic islands has long been recognized as representing a microcosm of the evolutionary process. As Darwin (1835) forecast shortly after his visit to the Galapagos Islands, ". . . the Zoology of Archipelagos will be well worth examining; for such facts would undermine the stability of Species." Biologists interested in the speciation process have been particularly impressed by the many clear-cut instances of adaptive radiation and of explosive speciation among insular forms (Lack, 1947; Grant, 1981; Givnish and Sytsma, 1997). Adaptive radiation is the process whereby a whole cluster of closely related but variously adapted taxa proliferates from a single common ancestor. (For a summary of definitions and discussion of adaptive radiation, see Givnish, 1997.) This phenomenon is a primary feature of speciation patterns in Hawaii and on oceanic islands in general. Typically, interisland founder events are followed by rapid expansions into a great variety of novel environments within each island, accompanied by differentiation and adaptation to the various ecological niches encountered there, generating new species. Islands evidently provide unique conditions promoting rapid speciation and adaptation to alternate environments. Chief among these factors is the requisite isolation that prevents ancestral genotypes from swamping diverging populations and that ensures restrictions on gene flow. Equally important are the ecological factors, particularly lack of competition in the early stages of colonization and a high level of environmental heterogeneity, the latter imposing differential selection on diverging populations in novel alternate habitats.

Every colonizing population passes through a genetic bottleneck (Nei *et al.*, 1975) that is more or less severe, depending on the number of founding individuals and the rate of growth of the founder population. At an extreme, just one propagule may be involved—a seed or an egg (a single genome), or alternately, a fertilized female (an insect, for example), that

brings the information of two diploid genomes (or more if multiply insem-
inated). In either case the founder carries a limited sample of the genetic
variability in the ancestral population. Without further contact with the
source population, the isolated founder population resulting from a successful chance introduction is free to diverge under the combined action
of the forces of random drift and selection. Intrinsic to all insular evolution
is isolation, engendered by both geographic and geological factors. Isolation is indubitably the key factor in evolutionary events on islands. Hence,
to understand the dramatic episodes of speciation and of evolutionary
diversification among Hawaiian organisms, it is important to first review the
geological history of Hawaii. With this backdrop, the isolating effects of volcanic factors can be appreciated on the broad scale, and on the smaller local
scale as well, in the discussion of several selected examples of speciation in
Hawaiian plants and animals that follows.

THE GEOLOGY OF HAWAII AND BIOLOGICAL CONSEQUENCES

The Hawaiian Islands are the youngest products of volcanic activity in
the earth's mantle at the Hawaiian hotspot, which is located near the middle
of the Pacific tectonic plate at about 19° N and 155° W (Clague and
Dalrymple, 1987) and some 4,000 km from the nearest major land mass. This
hotspot has been more or less continuously active for over 70 million years.
The accumulation of magma ascending in the mantle plume has generated
the more than 100 major volcanoes (Walker, 1990) strung out along a
6,000 km transect across the North Pacific as the Hawaiian–Emperor chain
(Fig. 1). Following their origin, emergence above sea level, and growth to a
high oceanic island, each volcano has been carried away from the hotspot
by the motion of the Pacific plate, which is currently moving northwestward
over the hotspot at a rate of about 9 cm/year. This ongoing volcanic activity has produced a "hotspot trace," an almost linear succession of islands of
increasing geological age drifting to the northwest in conveyer-belt fashion.
Over time, the processes of erosion and subsidence combined with the
development of coral reefs have reduced once-high islands to atolls. These
have further declined to ultimately sink below the ocean's surface and
become guyots or submerged seamounts, as are found in the western
Hawaiian Ridge and in the more northerly oriented Emperor Seamounts.
It is thought that the bend between the Hawaiian chain and the Emperor
chain west of Midway (Fig. 1) resulted from a dramatic shift in the direction of movement of the Pacific plate from north to northwest about
43 million years (Myr) ago.

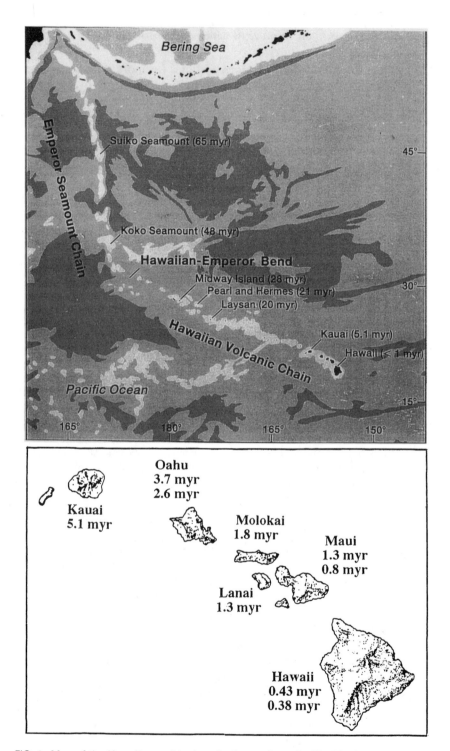

FIG. 1. Map of the Hawaiian archipelago in the northern Pacific. The inset shows the current high islands extending from Kauai to Hawaii.

This series of stages in the life cycle of a volcanic island is supported by potassium–argon radiometric datings that demonstrate the successively younger ages of volcanoes down the chain. The Emperor seamounts range in age from 70 to 43 Myr; the islands of the Hawaiian chain range in age from 28 Myr for Midway Island to the present-day lavas of the currently active Kilauea volcano on the Big Island of Hawaii, and the still submerged underwater volcano, Loihi, 30 miles to the southeast of Hawaii, which lies directly above the hotspot (Normark *et al.*, 1982). The northwestern eroded low islands and atolls of the Hawaiian chain are now depauperate in life forms but formerly must have had more diverse habitats and a richer biota when they were high islands (Carson and Clague, 1995). The current forested high islands are comparatively youthful in geological terms, with Kauai no more than 5.1 Myr old, Oahu 3.7 Myr old, the four islands of the Maui Nui complex—Molokai, Maui, Lanai, and Kahoolawe—from 1.9 to 0.8 Myr old, and the five volcanoes of the Big Island of Hawaii all no more than 0.43 Myr old (Macdonald and Abbott, 1970; Carson and Clague, 1995), with some still volcanically active. Of these, the massive shield volcano Mauna Loa is the largest on earth, rising nearly 9 km above the floor of the Pacific and 4,167 m above sea level.

With the range of altitudes on these younger high islands come sharp rainfall gradients and a dramatic diversity of habitats, encompassing coastal strands, arid lowland forests, moist rainforests, and even high elevation swamps and deserts. This rich array of habitats over small distances has been an important factor in promoting the adaptive radiation of the successful colonizers. However, the initial environments that greeted the first colonizers to the Hawaiian Islands were harsh and desolate, strewn with volcanic ash and lava from rift and summit eruptions. Ecological succession on the barren cooled lava flows beginning with lichens, mosses, and grasses slowly yielded soils capable of supporting shrubs and ultimately trees, with a succession of communities of plants and animals over the course of hundreds and thousands of years. Rarely would this be a straightforward process as continuing volcanic activity would repeatedly bury newly vegetated areas, locally extinguishing all plant and animal life, and laying the area bare again for subsequent recolonizations. That this is the normal course of events in Hawaiian environments is indicated by the frequent occurrence of "kipukas," remnant islands of vegetation spared from the course of the surrounding lava flows.

The instability of geologically young Hawaiian habitats is validated by radiocarbon dating of organic material recovered via drilling beneath prehistoric lava flows. On Mauna Loa on the island of Hawaii, studies show that despite its generally stated age of 0.4 Myr, over one half of Mauna Loa has been buried during the past 1,500 years and almost 90% of the surface is covered with rocks less than 4,000 years old (Lockwood and Lipman, 1987). The highly unstable nature of volcanic terrain has significant consequences for the evolution of colonizing populations; constant fragmentation of the habitat by lava flows generates small local population isolates that must be subject to the forces of random drift. Furthermore, repeated local extinctions and recolonizations imply multiple founding events throughout the range of a newly established species.

The variation observed between local populations of a species must be interpreted in the light of these volcanic factors that impose a series of isolating barriers between local populations and overall a metapopulation structure (Carson *et al.*, 1990). The evolution of the Hawaiian biota cannot be fully appreciated except in the context of both major and local volcanic events, and in addition, the Pleistocene changes in sea level (Stearns, 1966; Carson and Clague, 1995). Several times, rises in sea level have separated formerly contiguous populations on the islands of the Maui Nui complex. Whether by ocean channels or by inhospitable lava flows, Hawaiian populations of terrestrial organisms have frequently been subdivided by substantial isolating barriers, permitting independent evolution in the individual isolates.

SPECIATION IN HAWAIIAN PLANTS

Species endemism in Hawaiian flowering plants exceeds 90% (Carr, 1987). The current flora is interpreted as deriving from only 291 successful colonization events (Sakai *et al.*, 1995), of which about 10% were followed by spectacular adaptive radiations (Wagner, 1991). Four examples from three plant families are presented in the following sections. The most well-known radiation is that of the Hawaiian tarweeds, better known as the silversword alliance. This morphologically and ecologically diverse assemblage includes 28 species in three endemic genera, *Argyroxiphium*, *Dubautia*, and *Wilkesia* (Carr, 1985), which place in the tarweed subtribe (Madiinae) of the sunflower tribe (Heliantheae) of the family Asteraceae.

The Silversword Radiation

Silversword species have radiated into a diversity of terrestrial habitats in the Hawaiian Islands, from near sea level to elevations of 3,750 m, occupying localities with annual rainfalls less than 0.4 m to the wettest known on Earth with more than 12.3 m of rainfall. Adaptation to these diverse habitats and moisture regimes has been accompanied by amazing divergence in leaf and floral morphology, physiological tolerance, and most especially growth form. Habital forms include subherbaceous mats, cushion plants, rosette shrubs that flower once and die, decumbent and erect woody shrubs, large trees, and even vines, with the most extreme variation among the 21 species of the genus *Dubautia* (Carr *et al.*, 1989). Despite this remarkable anatomical and ecophysiological divergence, the group shows high genetic identity, as judged by isozyme data (Carr *et al.*, 1989) and by the frequent occurrence of natural hybrids even between genera (Carr, 1985). The group is clearly monophyletic. All the currently available data, including the molecular data (Baldwin *et al.*, 1990; Baldwin and Robichaux, 1995; Baldwin, 1997), support the hypothesis that the 28 Hawaiian tarweed species originated from a single colonizing ancestor, most likely carried to Hawaii in the form of a sticky fruit adhering to a migratory bird.

Although most Hawaiian plant introductions have been from Indo-Malaysian and Austral-Asian sources (Carlquist, 1970), evidence of several kinds has suggested a North American ancestor for this group; moreover, chloroplast DNA restriction-site comparisons have identified an endemic Californian genus, *Raillardiopsis*, which includes montane perennial herbs, as one likely ancestor (Baldwin *et al.*, 1991). This necessitates that the original founder underwent extreme long-distance dispersal across about 4,000 km of open ocean. A more serious problem for the hypothesis is that the ancestor and most of the Hawaiian Madiinae are self-incompatible, an exception to "Baker's rule" (Carr *et al.*, 1986) that plant founders from distant sources are generally self-fertilizing. Successful establishment of the tarweeds in Hawaii would thus have required simultaneous introduction of at least two founder individuals. The second ancestral individual could have been from the same or even from a different lineage. Interestingly, the Hawaiian silverswords, although showing normal diploid behavior at meiosis, are genetically tetraploid (Carr *et al.*, 1989), whereas *Raillardiopsis* is diploid.

The original Madiinae introduction to Hawaii must have occurred within the last 15 Myr, with the modern lineages of the silversword alliance estimated to have diverged 5.2 ± 0.8 Mya (Baldwin and Sanderson, 1998), the age of Kauai. Their current geographic distribution on all of the current high islands (Kauai to Hawaii, see Fig. 1) is consistent with an introduction on Kauai, or on one of the older now subsided islands (Carr *et al.*, 1989). Descent from a most recent common ancestor on Kauai is supported for the *Dubautia/Wilkesia* lineage by the molecular phylogeny derived from the 18–26S nuclear ribosomal DNA sequences of the internal transcribed spacer region (Baldwin, 1997). The origin of the apparent sister group *Argyroxiphium*, whose spectacular silversword and greensword species are restricted to Maui and to Hawaii, is, however, enigmatic. Following establishment, tarweeds dispersed to all the younger islands in the archipelago via at least 13 interisland founder events (Carr *et al.*, 1989; Baldwin and Robichaux, 1995), all but one of which are hypothesized to be from an older to a younger island. Perhaps as few as 5 of these 13 interisland founding events resulted in speciation (Baldwin and Robichaux, 1995).

The fact that the number of extant species on an island exceeds the number of hypothetical founders implies multiple intraisland speciation events, some of which may have involved intraisland founder events. This is particularly likely for the initial expansion of *Dubautia* from mesic into drier high-altitude habitats that can be inferred to have taken place within the Maui complex. This adaptive shift appears to have coincided with a chromosomal rearrangement that reduced the chromosome number from $n = 14$ to $n = 13$ (Carr and Kyhos, 1986). All nine derived species with $n = 13$ have a greater capacity than the twelve $n = 14$ *Dubautia* species to maintain turgor pressure as tissue water content decreases; this physiological difference is significant in their adaptation to harsher, drier environments (Carr *et al.*, 1989). Reconstruction of the historical pattern of ecological shifts on the molecular tree suggests that there were at least five ecological shifts from wet to dry habitats in the evolution of the silversword alliance, with at least three shifts to bog habitats (Baldwin and Robichaux, 1995). These major ecological shifts occurred on most or all islands and were apparently fundamental to speciation and diversification in the silversword alliance.

Radiation of the Genus *Bidens*

Adaptive radiation of the genus *Bidens* (Asteraceae) in Hawaii shows many features in common with the silversword radiation. The 19 endemic species and 8 subspecies of "beggar's ticks" are extremely variable morphologically and ecologically, much more so than in continental areas, and yet show little genetic differentiation based on the isozyme data (Helenurm and Ganders, 1985). All Hawaiian species of *Bidens* are completely interfertile (Ganders, 1989), another indication of their close genetic relationship, although hybrids are uncommon in nature because most species are allopatric. All the available evidence points to the fact that the diverse Hawaiian species evolved from a single common ancestor, most likely arriving in Hawaii via accidental long-distance dispersal by a migratory bird (Ganders, 1989). The barbed awns on the fruits of continental *Bidens* species are perfectly adapted for such a dispersal mechanism via birds or mammals. In Hawaii the lack of terrestrial mammals (except for one bat) together with the relative paucity of birds has led to an adaptive shift toward wind dispersal with many morphological modifications of the fruits, which may be flattened, winged, or helically twisted (Ganders, 1989).

By contrast with the silverswords, Hawaiian *Bidens* are self-compatible, facilitating successful colonization by a single immigrant; yet they include various mating systems promoting outcrossing, which undoubtedly contribute to the high levels of genetic variability found within Hawaiian populations (Helenurm and Ganders, 1985). Another factor is their polyploid status; all are ancient hexaploids showing diploid pairing at meiosis and $2n = 72$.

Species divergence in the Hawaiian taxa has been accompanied by many morphological modifications in addition to those of the fruits. Differentiation includes characters of growth habit (which ranges from prostrate herbs to small trees over 2 m); inflorescence structure; flowers and heads, which reflect adaptive shifts in pollination mechanism; and leaf shape, which ranges from always simple to bipinnately or tripinnately compound with numerous leaflets. Some of these morphological traits are correlated with habitat. Similarly to the silverswords, Hawaiian *Bidens* have expanded into a great diversity of habitats ranging from coastal bluffs and lithified sand dunes to cliffs, cinder cones, montane ridges, and bogs, at altitudes from sea level to over 2,200 m, with annual rainfall levels from 0.3 m to over 7.0 m. This extensive morphological and ecological differentiation may, however, have a relatively simple genetic basis and Ganders (1989) has presented preliminary data to suggest that only a few loci are involved.

The distribution of *Bidens* taxa on all the current high islands suggests an initial introduction on Kauai or on an older island. Speciation events have occurred in conjunction with both interisland founder events and with founder events within an island dispersing *Bidens* to other volcanoes or to other elevations. The two kinds of founder events have had variable outcomes: frequently differentiation of forms recognized as full species (19 cases), less often formation of subspecies (4 interisland and 4 intraisland), and in 11 interisland colonizations no recognizable divergence (Ganders, 1989). These facts emphasize that among plants, where genetic differences among insular taxa are minimal despite dramatic morphological and ecological differentiation, speciation is not an inevitable outcome of a founder event.

Radiation of the Lobelioid Genus *Cyanea*

The largest endemic genus of Hawaiian plants, *Cyanea* (family Campanulaceae), typically found in the understory in mesic and in wet forests at moderate elevations, has a woody habit and fleshy fruits that are dispersed by birds. The 55 named species of this genus, along with 39 members of three related lobelioid genera, *Clermontia*, *Rollandia*, and *Delissea*, are believed to have evolved from a single immigrant, making this radiation the most impressive among Hawaiian plants, at least in terms of species numbers (Givnish *et al.*, 1995, 1996). The genus *Cyanea* is the most diverse, being characterized by striking adaptive variation in growth form, in leaf size and shape, and in floral morphology. Species vary in height from 1 to 14 m and include treelets, shrubs and trees, and even one vinelike species. Many species exhibit developmental heterophylly, the juvenile leaves being more deeply divided than adult leaves, and 18 species have thornlike prickles (up to 1 cm in length) on their leaves, stems, and juvenile shoots. Both traits are unique for lobelioids, and the latter is particularly enigmatic, given the lack of native mammalian or reptilian browsers in Hawaii.

A molecular phylogeny based on cladistic analysis of chloroplast DNA restriction site variation (Givnish *et al.*, 1995) revealed two clades, a purple-fruited clade, and a larger orange-fruited clade, which also includes species of the genus *Rollandia*, now taxonomically submerged into *Cyanea*. Based on phylogenetic analysis of 40 populations from 35 (of the 63) species of *Cyanea–Rollandia*, it can be inferred that there have been at least 15 inter-island dispersal events, all but one following the typical pattern of dispersal down the island chain from older to younger islands. Twelve of the interisland founders have led to new species; an additional 22 intraisland speciation events have resulted from local isolation of populations, either by lava flows, by rises of sea level separating the islands of the Maui Nui complex, or via dispersal to volcanically disjunct areas on the same island. Despite or perhaps because of relatively low seed dispersal in *Cyanea* by frugivorous birds of the forest interior, radiation within an island has been more frequent than interisland dispersal.

The molecular phylogeny has also been used to analyze the evolution of the two unusual characters, prickles and heterophylly. The large prickles, found in 18 species of *Cyanea* and in 3 species of *Rollandia* (all in the orange-fruited clade), have evolved at least four times in the last 3.7 Myr. The juvenile–adult heterophylly trait largely parallels the prickle trait, having evolved at least three times, and both traits show increasing incidence on the younger islands (Givnish *et al.*, 1994). It has been hypothesized that these traits provided mechanical and visual defenses against herbivores, specifically against browsing by now extinct flightless geese and gooselike ducks (Givnish *et al.*, 1994) that were similarly distributed according to the subfossil remains but were exterminated by the Polynesians over the last 1,600 years (Olson and James, 1982). In this scenario, the potential adaptive value of these traits is clear, and Givnish *et al.* (1994, 1995) have made a convincing case for their hypothesis.

Radiation of the Alsinoideae

Yet another impressive adaptive radiation in the Hawaiian angiosperm flora, the fifth or sixth largest, is found in the subfamily Alsinoideae of the family Caryophyllaceae. The original transoceanic founder, perhaps from North America, probably colonized one of the older now eroded islands, and then descendants radiated throughout the archipelago to give rise to the 30 extant taxa: 4 species in the genus *Alsinidendron* and 25 species in the genus *Schiedea* with one species differentiated into two distinct varieties on separate islands (Wagner *et al.*, 1995; Sakai *et al.*, 1997). Speciation in the group has been accompanied by remarkable divergence in morphology (leaf shape, inflorescence structure, etc., see Fig. 2) and in habit (vines, compact or sprawling shrubs, perennial herbs), together with shifts in the breeding system and in the ecological habitat. Wagner *et al.* (1995) have used a phylogenetic analysis of morphological data along with character reconstruction to illuminate speciation patterns in the group; assuming extinction of ancestral taxa at each branching point, at least 56 speciation events can be identified. Four major clades, likely originating on Kauai, have undergone somewhat parallel diversification in terms of a succession of interisland colonizations from older to younger islands in the chain. A molecular phylogeny derived from chloroplast DNA and rDNA restriction site data (P. S. Soltis *et al.*, 1996) is largely congruent with the morphological phylogeny; combination of the morphological and molecular data sets yields a more robust phylogeny (Sakai *et al.*, 1997).

Despite the importance of interisland founders in the speciation and the geographic spread of this group, most speciations have been intraisland, with ecological differentiation being a major component. There have been repeated shifts from mesic to dry, often cliff habitats, concomitant with shifts from hermaphroditic to dimorphic breeding systems in the drier habitats, accompanied by the evolution of wind pollination. In certain cases there have been ecological reversals. Although the biogeographic pattern of colonization is indeterminate in parts of the phylogeny, perhaps due to extinct taxa, overall it seems that two thirds of the interisland colonizations (14/22) have resulted in speciation (Wagner *et al.*, 1995). It does not necessarily follow, however, that all the remaining cases correspond to instances in which a founder event failed to give rise to a new species. The current distribution of five taxa on two or more islands of the Maui Nui complex may not have required five or more interisland founder events. Rather, land

FIG. 2. Morphological diversity in habit, leaf, and inflorescence form in selected members of the four main clades of endemic Hawaiian Alsinoideae: *Schiedea membranacea* clade (A–C); *S. nuttallii* clade (D,E); *S. adamantis* (F); and *S. globosa* (G). The representatives of the first two clades are: A. *S. membranacea*; B. *Alsinidendron obovatum*; C. *S. verticillata*; D. *S. diffusa*; and E. *S. nuttallii* var. *nuttallii*. Reproduced with permission from Wagner *et al.* (1995).

bridges between the islands during periods of lower sea level in the Pleistocene may have permitted gradual range expansion to other volcanoes in the complex. Following the rise in sea level and separation of Maui, Molokai, Lanai, and Kahoolawe as individual islands, populations became allopatric isolates of the same species that have not subsequently undergone significant genetic or ecological differentiation.

Similar arguments can be applied to the multi-island Maui Nui distributions of *Bidens* species and species in the silversword alliance. However, where a species occupies two or more islands that have always been separated by deep ocean channels, a founder event without speciation is definitely implicated. Overall, however, a large proportion of plant interisland founders have led to the evolution of a recognizably new species, even if few genetic changes differentiate the ancestral and derived taxa.

SPECIATION IN HAWAIIAN ANIMALS

The extreme isolation of the Hawaiian Islands has severely restricted colonization by terrestrial vertebrates. There are no native amphibians or reptiles and only one mammal, the Hawaiian bat. Only birds have been successful in dispersing to Hawaii and radiating to generate a variety of landbirds that is considered spectacular for remote oceanic islands. Even so, the Hawaiian avifauna could be said to be depauperate, because only 6 of 174 extant families of songbirds worldwide are present in Hawaii (Freed *et al.*, 1987; Tarr and Fleischer, 1995). Nonetheless, it appears that there have been at least 20 avian colonizations that spawned over 100 species, the great majority of which are now unfortunately extinct (Olson and James, 1982; Freed *et al.*, 1987).

Among invertebrates, the most prolific groups in Hawaii are the land snails, which include some 752 extant species from only 10 of the world's 70 to 100 land snail families (Cowie, 1995), and even more impressive, the insects with perhaps 10,000 native species from 16 orders (Nishida, 1994, 1997; Miller and Eldredge, 1996), with many species still to be named. Hawaiian insects include representatives of 15% of the world's nearly 1,000 insect families (Simon, 1987). The spectacular insect radiations in Hawaii stemmed from rare founding events following long distance passive dispersal. Insects could have been transported to Hawaii as adults carried by storms or via the high altitude jet streams, or alternately (Otte, 1989), as eggs deposited in floating vegetation. To account for the Hawaiian insect fauna only 350 to 400 separate colonizations are required with only one successful arrival every 75,000 to 175,000 years (Howarth and Mull,

1992). The two most speciose Hawaiian insect groups are the beetles (order Coleoptera) with some 3,000 species and the true flies (Diptera) with over 1,500 species, followed by the moths (Lepidoptera) and the wasps and bees (Hymenoptera), each of which includes almost 1,000 species. Here, three insect examples are discussed: the crickets, the planthoppers, and the drosophilid flies. These groups provide more insight into the nature of species barriers and into the levels of pre- and postzygotic isolation between insular taxa than can be gleaned from the data available for Hawaiian birds.

Speciation in Hawaiian Honeycreepers

The most predominant and diverse group of Hawaiian passerine birds are the "honeycreepers" (Amadon, 1950) or Hawaiian finches (subfamily Drepanidinae in Fringillidae). Some 50 species have now been described (17 only from subfossil remains), and all are thought to have evolved from a single founder that was most likely a primitive cardueline finch with a small generalized bill (Raikow, 1976). Monophyly of the group is strongly supported by the allozyme data of Johnson *et al.* (1989) and by the most recent molecular phylogeny based on sequences of the mitochondrial (mt) *cytochrome b* gene (Fleischer *et al.*, 1998). Descendants of the original founder have undergone extreme differentiation in diet, in bill and tongue morphology, in plumage coloration, and in behavior, diverging significantly from mainland forms. The three recognized tribes, the Drepanidini ("red" birds), the Psittirostrini ("finch-billed" birds) and the Hemignathini ("green" birds) include nectivorous, insectivorous, frugivorous, granivorous, and omnivorous forms. Bills vary remarkably (Fig. 3) in size, thickness, length, and shape (straight or variously decurved), each specialized for its particular food resource (Freed *et al.*, 1987). The nectar sources provided by the tubular flowers of *Cyanea* and by other lobelioids may well have been important in the evolution of the long-billed forms. Interestingly, the adaptive shift to nectar feeding also entailed the evolution of a tubular tongue, together with associated anatomical specializations, specifically a nasal operculum that presumably prevents pollen from entering the nasal cavity (Raikow, 1976).

The extensive morphological diversity accompanying the adaptive radiation of the Hawaiian honeycreepers has been achieved without a comparable degree of genetic differentiation, as evaluated by restriction-site analyses of mtDNA (Tarr and Fleischer, 1993, 1995). Moreover, their close genetic relationships support rapid divergence of the honeycreepers in Hawaii. Using the geological information to calibrate the rate of mtDNA

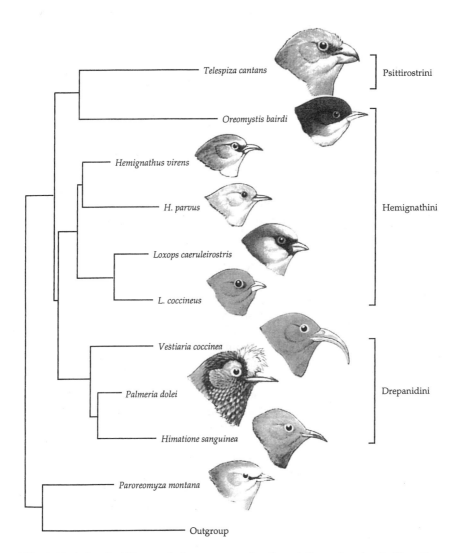

FIG. 3. Variation in bill morphology among the three tribes of endemic Hawaiian honeycreepers mapped onto the molecular phylogeny of Tarr and Fleischer (1995) inferred from mtDNA restriction site variation. Reproduced from the article of Givnish (1997) in *Molecular Evolution and Adaptive Radiation* edited by T. J. Givnish and K. J. Sytsma (ISBN 0-521-57329-7), with the permission of Cambridge University Press.

sequence divergence and extrapolating back to the base of the Hawaiian honeycreeper tree, Fleischer *et al.* (1998) inferred that the basal split into the two main clades of creepers (e.g., *Paroreomyza montana* in Fig. 3[1]) versus the remaining drepanidines occurred 4–5 Mya, approximately the age of Kauai. Hence, the quite extraordinary morphological and ecological divergence in the group has occurred in a relatively brief time span, geologically speaking. Because some honeycreepers function as pollinators and dispersers of *Cyanea* and other lobelioids, these lineages must have coevolved with the lobelioids over the past few million years. Although Givnish *et al.* (1995) suggested that this association may extend back in time to occupation of older, now subsided islands, the data of Fleischer *et al.* (1998) indicate that this idea may need to be reevaluated. Regardless of their times of origin, the fact that 22% of all *Cyanea* species are now extinct may be due in part to historical extinction of their honeycreeper pollinators.

In widespread and well-differentiated lineages such as the amakihis (represented in Fig. 3 by *Hemignathus virens*), sister taxa are found on different islands, with differentiation presumably a consequence of interisland dispersal and of founder events. Although currently considered to be subspecies, some island forms of the amakihi complex may deserve species status (Tarr and Fleischer, 1993). Speciation or subspeciation has not always followed interisland dispersal, however. A number of honeycreepers, such as the nectar-feeding i'iwi *Vestiaria coccinea* and apapane *Himatione sanguinea sanguinea* (see Fig. 3), have multi-island distributions (Freed *et al.*, 1987). This is not surprising given the high vagility of the nectivorous and frugivorous honeycreepers, which forage widely, flying over the canopy in search of food. They may occasionally fly actively or be carried passively by wind between islands, effecting a measurable level of gene flow between island populations. In other species, their much more restricted flight patterns facilitate greater interisland differentiation. It is important to note here that behavioral parameters (in this case foraging behavior) may have a significant effect on speciation patterns in animals. As genetically determined traits, behaviors evolve, leading to behavioral differentiation, which may further compound the morphological and ecological differentiation so typical of insular speciation.

---

[1] It should be noted that the honeycreeper phylogeny based on *cyt b* sequences presented in Fleischer *et al.* (1998) differs in a number of details from the preliminary phylogeny of Tarr and Fleischer (1995) based on restriction site data, and shown in Fig. 3; for example the Kauai creeper *Oreomystis bairdii* and the Maui creeper *P. montana* now cluster together as a separate clade. The more recent molecular phylogeny is in better accord with relationships inferred from other types of data (e.g., Johnson *et al.*, 1989) and hence is considered more reliable.

Speciation in Hawaiian Crickets

One of the successful arthropod groups in Hawaii is the family Gryllidae (order Orthoptera) that has diversified into some 240 species in 7 genera. These endemic species derive from three separate introductions of a tree cricket, a sword-tail cricket, and a ground cricket, respectively (Otte, 1989; Shaw, 1995). (Another oceanic ground cricket that is widespread throughout the Pacific has not speciated in Hawaii.) All Hawaiian crickets are flightless and are endemic to single islands, and within-island speciation is the predominant pattern (Otte, 1989; Shaw, 1995, 1996b). Some speciation events have been associated with adaptive shifts, most notably in the genus *Caconemobius* (subfamily Nemobiinae). The ten species in this genus of ground crickets derive from a marine shoreline-inhabiting ancestor that probably rafted to Hawaii from the western Pacific; these beach crickets then likewise dispersed to the shores of all islands in the archipelago. An adaptive shift to freshwater habitats has occurred independently on at least four islands (Howarth and Mull, 1992) as the shore species repeatedly invaded inland habitats.

The fascinating radiation on the youngest island of Hawaii has yielded a black, surface-living species *C. fori* that forages by night on the very young, as yet unvegetated lava flows, and subsequently at least three cave species that have adapted to subterranean life in lava tubes, probably within the past 100,000 years. These remarkable cave species on Hawaii, and the two species that independently evolved on Maui from a parallel adaptive shift, display all the characteristic cave adaptations of reduced pigment, small eyes, and translucent exoskeleton (Howarth, 1972; Howarth and Mull, 1992). The colonization of individual lava tubes likely involves independent founder events, and consistently, populations in different lava tubes are distinct, although they are purportedly the same species. On the older islands the lack of similar pale, blind cricket species is probably due to loss of these specialized habitats once volcanoes became extinct. By contrast with most Hawaiian crickets, *Caconemobius* species are mute and males cannot attract females via acoustic signaling.

The tree crickets (subfamily Oecanthinae) have radiated into three endemic Hawaiian genera with 4, 24, and 36 species, respectively, that comprise 43% of the world's known tree cricket species (Otte, 1989). This substantial radiation from a single American ancestor has involved considerable morphological and ecological diversification with adaptive shifts from trees and bushes to ground dwelling and even into the subterranean habitat. In a phylogenetic analysis of 28 of the 36 species of *Prognathogryllus* based on morphological traits, Shaw (1995) demonstrated a

Kauai origin with sequential evolution to Oahu and the younger islands, but with intraisland speciation being the predominant mode.

The forest-inhabiting sword-tail crickets (subfamily Trigonidiinae) have undergone the largest radiation to form 166 species placed in three genera (Otte, 1989; Shaw, 1995). This group occupies fewer adaptive zones and species are for the most part morphologically cryptic. Although flightless, the forewings are retained and modified to function as sound-producing organs. Species are reproductively differentiated by a behavioral trait, the male calling song, which has undergone rapid evolutionary divergence (Otte, 1989). Sexually mature females respond specifically to the songs of males of their own species, these acoustic signals serving for mate localization as well as for species recognition and discrimination among sympatric species. The primary reproductive barriers among these and many other cricket species are thus prezygotic, and a new species can conceivably arise from the evolution of a distinctive male call coupled with the evolution of female preference for that call. In fact, based on his analyses of Hawaiian crickets, Otte (1989) contended that these behavioral reproductive barriers evolve first, long before any ecological or morphological differentiation, and furthermore, that speciation can occur in continuous forest before major geographic barriers arise.

Cricket calls can be quite complex, varying in a number of parameters. Nonetheless, among the 36 species of the endemic Hawaiian sword-tail genus *Laupala*, call variation is simple, differing only in pulse rate. Sympatric species have distinct pulse periods and the difference is genetically determined. Shaw (1996a) has demonstrated the polygenic basis of this behavioral trait in crosses between two closely related but acoustically well-differentiated species, *L. kohalensis* and *L. paranigra*, both from the island of Hawaii (Fig. 4). The data suggest a Type I genetic architecture (i.e., many genes, each of small effect—Templeton, 1981); at least 10 genetic factors are estimated to underlie the interspecific difference, with the X chromosome making a nonadditive contribution to the phenotype (Shaw, 1996a).

Male calls can apparently diverge rapidly as indicated by the intraspecific difference between *L. paranigra* populations on different volcanoes (Shaw, 1996a), and small interpopulational differences can presumably gradually accumulate to achieve full species differentiation. It should be noted that the success of these laboratory hybridization experiments points out the lack of postmating barriers between these sister taxa of *Laupala*. Moreover, although generally effective in nature, the premating barriers can be readily overcome in the laboratory, facilitating genetic analysis of species differences. In a few rare cases premating barriers are apparently incomplete, permitting localized zones of hybridization in the field (Otte, 1989). Most often however, sympatric species are reproductively distinct,

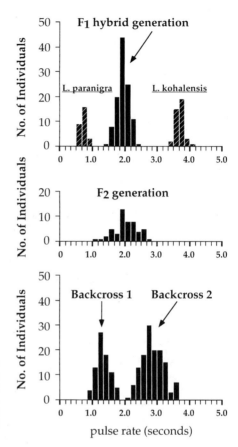

FIG. 4. Evidence for the polygenic basis of interspecific variation in pulse rate in the songs of two endemic Hawaiian crickets, *Laupala paranigra* and *L. kohalensis*. Reproduced from Shaw (1996a) with permission from The Society for the Study of Evolution.

well isolated by a simple behavioral barrier that may have a relatively complex genetic basis.

As in other Hawaiian cricket genera, speciation events in the genus *Laupala* have been mainly intraisland, but speciation has also resulted from each of the rare interisland founders implicated in the sequential spread of the genus from Kauai to Oahu and thence to Maui and to Hawaii (Shaw, 1996b). In addition, the phylogeny inferred from mtDNA sequences of 17 of the 36 species in the genus indicates two back migrations from Hawaii to Maui, with speciation following from these founder events as well (Shaw, 1996b). Although the pattern suggested by these molecular data is not in complete agreement with the taxonomic hypothesis based on morphology and calling songs, the mtDNA data confirm that individual *Laupala* species are endemic to single islands and are most closely related to other species from the same island. Descendants of each successful interisland founder have thus radiated within each island, the male calls diverging to occupy the bioacoustic space, and the genes encoding this reproductive character diverging along with other components of the genome.

Speciation in Hawaiian Planthoppers

The sap-feeding planthoppers (Homoptera: Fulgoroidea) are ubiquitous in virtually all Hawaiian terrestrial ecosystems. Of the 18 families worldwide, only two families, the Cixiidae and the Delphacidae, have colonized Hawaii to generate at least 206 endemic species and 27 subspecies (Asche, 1997). Major radiations have occurred in the cixiid genus *Oliarus* with 58 species and 24 subspecies currently recognized, and in the delphacid genus *Nesosydne* that includes 82 Hawaiian species and one subspecies. Based on shared morphological characters of tarsal spines and male genital characters, the genus *Oliarus* is considered monophyletic, possibly deriving from a North American ancestor. The morphologically heterogeneous *Nesosydne*, on the other hand, appears to be polyphyletic and descended from multiple founders (Asche, 1997).

Most Hawaiian planthoppers are monophagous, with over 70 Hawaiian plant genera reported as hosts. These include tree ferns, grasses, palms, herbs, shrubs, and trees. Apart from these data on host–plant associations, relatively little is known of the biology of Hawaiian planthoppers, the group having been rather neglected up until recently. Among planthoppers in general, morphological divergence is limited, mate recognition depending to a large extent on distinctive acoustic signals (Claridge, 1985). These consist of trains of damped pulses produced by specialized structures, called tymbals, located bilaterally in the first two abdominal segments. Sounds are produced by both sexes and are substrate-borne, with the host plant serving to transmit these low-frequency vibrations from one individual to another (Claridge, 1985). Thus speciation in planthoppers usually involves a shift in the specific mate recognition system (Den Hollander, 1995), with reproductive isolation between species effected at the prezygotic level. Behavioral divergence may be accompanied by ecological divergence, resulting from either habitat or host plant shifts. Although data are limited, these various factors are also implicated in the speciation of Hawaiian planthoppers. In the genus *Oliarus*, studies have focused on the acoustic divergence distinguishing populations in a unique habitat (Hoch and Howarth, 1993), whereas in *Nesosydne* the focus has been on host plant associations of planthopper species in one section of the genus (Roderick and Metz, 1997).

In the adaptive radiation of the genus *Oliarus*, there have been at least six independent shifts from surface to cave habitats, with one invasion of lava tubes on the island of Molokai, three on Maui, and at least two on the Big Island of Hawaii. Natural selection on planthoppers in the cavernicolous habitat has resulted in the morphological modifications typical of obligate troglobites, namely, reduction of eyes, wings, and pigment (Howarth, 1983, 1993). This habitat shift has also imposed constraints on the behavioral communication systems essential to successful reproduction (Hoch and Howarth, 1993). Whereas mate recognition in surface-dwelling *Oliarus* species probably includes a visual component, judging by the large compound eyes of epigean species, visual cues are precluded by the permanent darkness of the cave habitat and the completely eyeless state of the cavernicolous *Oliarus*. Communication and species recognition apparently depend solely on the vibrational signals they produce that are efficiently transmitted over distances of a few meters along the dangling root strings that penetrate the cave habitat. These roots of *Metrosideros polymorpha* and of other native Hawaiian trees that successfully colonize the young lava flows above provide an abundant food resource for *Oliarus* nymphs and adults, as well as their communication medium.

Courtship in the cave species *Oliarus polyphemus* from the island of Hawaii is typically initiated by a receptive female, with a conspecific male responding to a calling stationary female by moving toward her, calling occasionally, until he locates her and initiates copulation about an hour later, at which point both sexes stop calling (Hoch and Howarth, 1993). In contrast with crickets where only the male calls, in planthoppers both sexes function as "senders" and as "receivers" of species-specific acoustic signals, the courting couple in essence singing a duet (Howarth *et al.*, 1990). Whereas in epigean planthoppers calling is usually initiated by a male, in the cavernicolous *Oliarus* the behavioral pattern is switched, with the female both initiating calling and calling much more actively than the male. Hoch and Howarth (1993) speculated that this behavioral switch may be an additional adaptation to the cave environment, making mate location more economical.

Remarkably, significant acoustic differentiation has been found among populations of *O. polyphemus* from different lava tubes. Between caves, courtship signals differ in call duration and in interpulse interval in each sex, but with only minimal differences between calls of males and females from the same cave (Hoch and Howarth, 1993). In playback experiments males show high preference for calls of females from their own population, supporting the hypothesis that these signals provide effective premating isolation. Although there is no detectable ecological or morphological divergence among *Oliarus* from caves that are separated by 20 km at most, the seven populations analyzed are considered reproductively distinct, being regarded as cryptic acoustical species (Howarth *et al.*, 1990; Hoch and Howarth, 1993). The acoustic divergence is believed to have occurred rapidly, perhaps within as little as 60 to 100 years, the age of some of the occupied lava tubes, following dispersal to new sites through the subterranean system. Hoch and Howarth (1993) argued for allopatric behavioral divergence among disjunct lava tubes, following the initial shift to the cave environment by a single ancestor derived from a rainforest species. They explained the coexistence of two acoustic forms in one cave as a case of secondary overlap.

In the large and polyphyletic Hawaiian genus *Nesosydne*, planthopper species feed on a wide diversity of hosts, being recorded from 28 plant families, although individual species show high host plant specificity (Roderick, 1997). One cluster of about 15 *Nesosydne* species, which may represent a monophyletic clade, is restricted to the Hawaiian silverswords, planthoppers of this group having been collected on 13 of the 28 members of this alliance (Roderick and Metz, 1997). Each planthopper species is highly host-specific, being associated with a single plant species, or in a few cases two very closely related plants and their hybrids. Preliminary phylogenetic studies indicate that diversification in this group of *Nesosydne* planthoppers parallels diversification in the silversword alliance, with a significant number of cospeciation events between these planthoppers and their hosts (Roderick and Metz, 1997).

Cospeciation in the two lineages has been documented by comparing a partial planthopper molecular phylogeny based on nucleotide sequences of the mt *cytochrome oxidase I* gene (Roderick and Metz, 1997) with a portion of the silversword molecular phylogeny based on nuclear rDNA sequences (Baldwin and Robichaux, 1995) and testing the significance of the match between the two phylogenetic patterns (Page, 1995). The six *Nesosydne* species analyzed from this complex feed on eight silversword species; two of the planthopper species, *N. naenae* and *N. chambersi*, each feed on two closely related *Dubautia* species (as well as their hybrids), whereas the remaining four each have a unique silversword host (Fig. 5). Comparison of the two phylogenies shows that they are broadly congruent, each being monophyletic and comprising three main clades, with five cospeciation events identified among these planthoppers and their silversword hosts (indicated by **a** to **e** in Fig. 5). Randomization tests indicate that this level of cospeciation is statistically significant ($p < 0.01$; Roderick and Metz, 1997). It will be critical to determine whether the match between the phylogenetic patterns of these insects and their host plants is further substantiated when the planthopper phylogeny is expanded to include the remaining nine species of this *Nesosydne* complex.

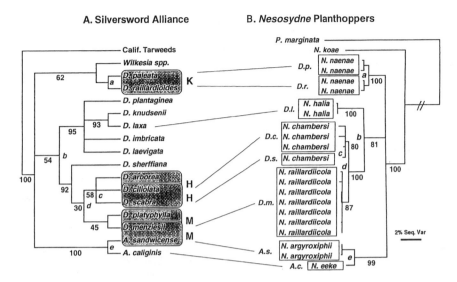

FIG. 5. Comparison of portions of the phylogenies of members of the silversword alliance (**A**) and the delphacid *Nesosydne* planthoppers (**B**), suggesting significant cospeciation between these herbivorous insects and their host plants (points *a* to *e* in the two phylogenies). The host associations of planthopper species are shown by the lines interconnecting **A** and **B**. The silversword phylogeny shown (derived from that of Baldwin & Robichaux, 1995) includes 13 species of *Dubautia* (*D.*) and two species of *Argyroxiphium* (*A.*) endemic to Kauai (K), Maui (M) and Hawaii (H), with species that hybridize indicated by the shaded ovoid boxes. The *Nesosydne* phylogeny includes 6 species, with sequenced individuals using the same host plant boxed together, along with one Hawaiian (*N. koae*) and one non-Hawaiian delphacid outgroup species (*P. marginata*). Numbers below the branches are bootstrap percentages indicating the level of statistical support for that branch in the phylogeny. Reproduced from Roderick (1997) with permission from the University of Hawaii Press.

The shared patterns observed to date support an interpretation of cospeciation of these planthoppers with their silversword hosts, but they do not reveal the underlying processes or discriminate among the several possible mechanisms that may be responsible. There is no evidence for host plant switching. Rather, it appears that this planthopper radiation followed the silversword radiation, although it is yet to be demonstrated whether this occurred via coevolution and specific adaptation of these planthopper species to their respective host plants (Roderick and Metz, 1997; Roderick, 1997). It is possible that this complex of *Nesosydne* species derived from a single founder that successfully colonized an ancestral silversword species on Kauai shortly after its establishment ~5 Mya, the group subsequently proliferating as it spread down the archipelago to Hawaii, diversifying along with its silversword hosts. Unfortunately, there are no published data on acoustic signaling in these delphacid planthoppers to determine whether sexual selection has also played a role in their speciation processes, in conjunction with the obvious ecological component.

Speciation in Hawaiian *Drosophila*

Of all the impressive radiations among Hawaiian organisms, none shows such explosive speciation as the endemic Drosophilidae (order Diptera). From one or perhaps two original founders (Throckmorton, 1966), close to 1,000 species have evolved. Hawaiian species represent at least one fourth of the world's known drosophilid fauna, an amazing proportion given Hawaii's restricted land area of 6,318 square miles, only a small portion of which is suitable forest. Taxonomically, the majority of these flies fall within one or other of two main genera, *Drosophila* and *Scaptomyza* (Kaneshiro, 1976a; Hardy and Kaneshiro, 1981), with estimates of at least 600 and 260 species, respectively. The two lineages are morphologically and behaviorally distinct, but similarities in internal anatomy suggest their close relatedness and the possibility that the drosophiloids and scaptomyzoids arose from a single introduction (Throckmorton, 1966). The startling implication of this scenario is that the genus *Scaptomyza* originated in Hawaii and then dispersed from Hawaii to other parts of the world.

Hawaiian drosophilids display quite remarkable morphological, behavioral, and ecological modifications unknown in continental *Drosophila* (Carson *et al.*, 1970; Carson and Kaneshiro, 1976). In size they range from some of the smallest drosophilids known (ca. 1.5 mm) to giant flies with a wingspread of up to 18 to 20 mm (Hardy, 1965). Many of the unique morphological traits are sexually dimorphic and represent secondary sexual characters that are involved in the elaborate male courtship behaviors (Spieth, 1966, 1974, 1982). These prominent morphological features are the basis for various convenient groupings within the Hawaiian drosophiloids (Hardy and Kaneshiro, 1981), such as the picture-wings (111 species), modified-mouthparts (ca. 100 species), ciliated-tarsi, bristle-tarsi, split-tarsi, spoon-tarsi, and white-tipped scutellum groups. Behaviorally, Hawaiian drosophiloids are unique in their complex and diverse species-specific courtship patterns, in the male defense of lek territories, and their much higher level of agonistic behavior compared to non-Hawaiians (Spieth, 1974). In the scaptomyzoid lineage, by contrast, behavior is simpler, with assault-type mating and no territorial defense.

In addition to their extraordinary morphological and behavioral diversity, Hawaiian drosophilids are ecologically diverse, having adapted to both wet and to dry montane forests at a range of altitudes, and having radiated to utilize a wide variety of breeding substrates. These saprophagous flies lay their eggs in various parts of decaying plants or fungi, leading to informal groupings such as leaf-breeders, bark-breeders, flower-breeders, flux-breeders, and fungus-breeders (Heed, 1968, 1971). Other substrates include rotting fruits and exposed roots, and some species even oviposit in spider's eggs (Hardy, 1965). Some 40 families of Hawaiian plants have been recorded as breeding substrates (Heed, 1968; Montgomery, 1975; Carson and Kaneshiro, 1976), and although the majority of drosophilids (75%) are strictly monophagous, the degree of host plant species specificity varies to include some flies that are polyphagous on five or more host plant families (Kambysellis and Craddock, 1997). Typically, adult females are extraordinarily selective in their choice of an oviposition site. This behavior appears to be a genetically determined trait (Ohta, 1989) that underwent rapid divergence early in the colonization and adaptive radiation of drosophilids in Hawaii; in fact, their remarkable ecological divergence may have been directly instrumental in the initial phyletic diversification of this group in Hawaii (Kambysellis et al., 1995).

The close adaptation of these flies to their specific ecological niches extends to anatomical and developmental modifications of the female reproductive system (Kambysellis and Heed, 1971; Kambysellis and Craddock, 1997; Craddock and Kambysellis, 1997), with consequent shifts in their reproductive strategies, in addition to the shifts in oviposition behavior. Wide differences in ovarian structure and functioning achieve quite varied fecundity potentials. At one extreme, the flower-breeders have as few as two ovarioles per fly, with alternating ovariole development, so that at most only one egg can be matured and deposited. At the other extreme, the bark-breeders can oviposit hundreds of eggs at a time as a result of large numbers of ovarioles per fly (greater than 80 in some species), synchronous ovariole development, and up to three mature eggs per ovariole. The size of the egg mass produced correlates with the nutritional reserves available in the various substrates, optimizing larval survival in each niche and the ecological adaptation of each species (Kambysellis and Heed, 1971). In large measure, the size of the egg mass is varied by species-specific regulation of the expression rate of the three vitellogenin genes that encode the yolk proteins, the major constituents of the egg. These proteins vary interspecifically in size as well as in quantity (Craddock and Kambysellis, 1990), reflecting the occurrence of insertion/deletion mutations in coding sequences of the vitellogenin genes (Ho et al., 1996) as well as base substitutions (Kambysellis et al., 1995) in coding and regulatory regions.

Further adaptation to the microenvironment of the breeding niche is suggested by striking variations in the morphology of the egg; the respiratory filaments display different lengths and degrees of porosity, and chorion ultrastructure varies adaptively with dramatic divergence in all the regions involved in respiratory exchange and in the thickness and complexity of the outer endochorion (Fig. 6; Kambysellis, 1993). The length and shape of the ovipositor also vary in accord with the nature of the oviposition substrate (Craddock and Kambysellis, 1997). Moreover, female sexual behavior appears to be adaptive in that female receptivity to male courtship by and large correlates with the physiological maturation of their ovaries; by sexual maturity there are no unmated females in natural populations (Kambysellis and Craddock, 1991; Craddock and Dominey, 1998).

The spectacular variety and fantastic number of Hawaiian *Drosophila* species naturally lead us to question the nature of the underlying speciation events and the nature of the reproductive barriers that separate closely related species. These barriers can be assayed in hybridization tests wherever laboratory culture is possible. This largely restricts analysis to the well-known picture-winged species, many of the other groups being difficult or impossible to culture. From a review of the then available hybridization data, Craddock (1974a) concluded that reproductive relationships between Hawaiian *Drosophila* species are highly varied, reflecting various degrees of phylogenetic relationship, different rates of species differentiation, and different speciation modes. Typically, sympatric species are separated by prezygotic barriers resulting from differences in male courtship behavior that normally preclude interspecific matings. Allopatric species mate more readily but are separated by postzygotic barriers, most often hybrid male sterility, and rarely hybrid inviability (Craddock, 1974a,b). Early stages in the evolution of the species gap can best be monitored by focusing on closely related and recently evolved forms, as exemplified by the following instances of allopatric and sympatric taxa.

Differentiation in the Drosophila grimshawi Complex

Within a species subgroup, related allopatric taxa from different Hawaiian islands are almost invariably recognized as distinct species, the

→

FIG. 6. Variation in eggshell thickness and structural complexity among Hawaiian and non-Hawaiian members of the genera *Drosophila* and *Scaptomyza*. Below each scanning electron micrograph (SEM) is an interpretive drawing of the outer endochorion cross-section. The SEMs are taken from representative species as follows, from left to right: *Scaptomyza albovittata*, *D. (Hirtodrosophila) pictiventris*, *D. (Drosophila) adunca*, *D. (Drosophila) dolichotarsus*, *D. (Drosophila) disjuncta*. Note that whereas the scaptomyzoids are the most primitive of the Hawaiian flies and the picture-winged group the most derived, this diagram does not accurately portray the phylogenetic arrangement of the intermediate groups of flies (see Fig. 8). Reprinted from *Int. J. Insect Morphol. & Embryol.*, Vol. 22, Kambysellis, M. P./ Ultrastructural diversity in the egg chorion of Hawaiian *Drosophila* and *Scaptomyza*: Ecological and phylogenetic considerations, Fig. 18, Copyright (1993), with permission from Elsevier Science.

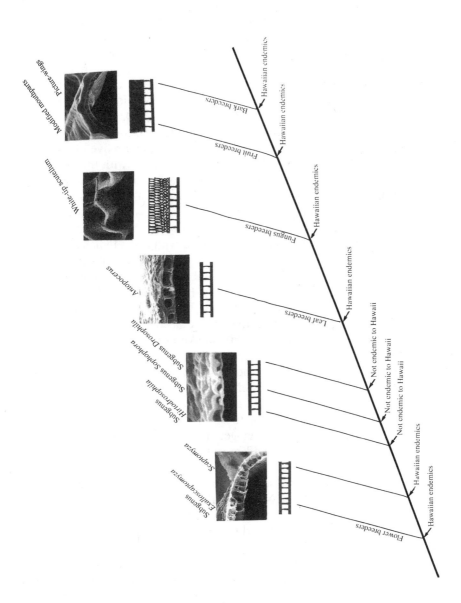

vast majority of Hawaiian *Drosophila* being single-island endemics. The multi-island distribution of *Drosophila grimshawi* presents a conspicuous exception in that several interisland founder events appear to have failed to trigger the usual outcome, that is, speciation, with substantial genetic and morphological differentiation. As the "standard" picture-winged species for cytological studies (Carson, 1970, 1983, 1992) and a species that is easy to culture, *D. grimshawi* has been widely used for developmental (e.g., Dickinson, 1980; Kambysellis *et al.*, 1986, 1989), behavioral (e.g., Ringo, 1976; Ringo and Hodosh, 1978; Droney, 1992, 1994; Droney and Hock, 1998), and molecular studies (e.g., Martinez-Cruzado *et al.*, 1988; Hatzopoulos *et al.*, 1989). Yet its anomalous distribution on five of the high islands (Kauai, Oahu, Molokai, Maui, and Lanai), along with the apparent morphological uniformity of these island populations, is quite atypical for Hawaiian flies. A related population from the youngest island, Hawaii, that was previously included in the species *D. grimshawi*, is now recognized as a separate species, *D. pullipes*, on the basis of a minor color difference and the production of sterile hybrids when crossed with Maui females.

The evolutionary status of members of this intrinsically interesting complex of island taxa has been addressed via interisland hybridization studies. Some combinations of *D. grimshawi* (e.g., Kauai × Oahu, and Maui × Molokai) show no evidence of hybrid breakdown, confirming their genetic compatibility and status as members of a common species, whereas others (e.g., Maui × Oahu) show fertility breakdown in the F_2 and backcross males, although they show full fertility in the F_1 males (Ohta, 1980). This partial postmating isolation, along with evidence of some premating isolation (Ohta, 1978) and other behavioral (Ringo, 1976) and ecological differentiation (Montgomery, 1975), all features of species divergence, indicated that further examination of these taxa was in order. Accordingly, Piano *et al.* (1997) undertook a phylogenetic analysis of the complex, using both molecular sequence data and morphological data (ultrastructure of the chorion), as well as all the previously available data. Analyses of the combined and individual data sets all concurred in indicating two clades— one that includes all the Maui Nui populations, and the other comprised of the Kauai and Oahu populations along with *D. pullipes* of Hawaii.

The major distinction between the two clades is ecological. Females of the Kauai, Oahu, and Hawaii populations are ecological specialists, ovipositing exclusively in the decaying bark of *Wikstroemia* (family Thymeleaceae); the Maui Nui taxa are ecological generalists, using bark of more than five plant families for breeding (Montgomery, 1975). The evidence of additional genetic differentiation coincident with the ecological differentiation indicates that the two forms of *D. grimshawi* are more properly considered as full species rather than as incipient species, although they are morphologically cryptic except for the different chorionic patterns. The

Kauai and Oahu populations have recently been named as a new species (Kaneshiro and Kambysellis, 1999), whereas the Maui Nui populations will retain the name *D. grimshawi* as originally designated, making three species in this complex of closely related forms.

Given that monophagy is the ancestral state for the Hawaiian *Drosophila* (Kambysellis *et al.*, 1995), it can be inferred that following the colonization of Maui Nui by an ecologically specialist ancestor from Kauai or perhaps an even older island, the founder population underwent an expansion of host plant use. This quite dramatic ecological and behavioral shift accompanying the speciation of *D. grimshawi sensu stricto* involved minimal morphological change and perhaps occurred in response to competition for breeding substrates. As the most derived of all the Hawaiian Drosophilidae (Kambysellis *et al.*, 1995), it is likely that colonizers of the *grimshawi* species group found their favored breeding niches already occupied and were therefore forced to adapt to new substrates and to broaden their host plant use or perish. As currently interpreted by the phylogenetic hypothesis of Piano *et al.* (1997), three founder events in the history of the *D. grimshawi* complex resulted in speciation, as is the norm for Hawaiian *Drosophila*. In the ecologically specialist clade, a founder event on Kauai led to the newly recognized species, *Drosophila craddockae* (Kaneshiro and Kambysellis, 1999) and a second founder event on Hawaii by a Kauai ancestor led to *D. pullipes*; in the sister clade a founder reaching one of the islands of the Maui Nui complex gave rise to the ecologically generalist species *D. grimshawi* that subsequently expanded to the adjacent islands during times of lower sea level. Accepting this scenario, the *D. grimshawi* complex is not quite so aberrant; the only significant anomaly is that the population on Oahu founded by a Kauai ancestor failed to undergo any substantial genetic differentiation and speciate following the founder event. Interestingly, a parallel case occurs in *Drosophila crucigera*, another ecologically generalist species of the *grimshawi* species group that also inhabits the islands of Kauai and Oahu. These exceptions among allopatric taxa emphasize that just as in plants, in animals speciation is not always an automatic sequel to a founder event, although it is certainly the most frequent outcome.

Differentiation between Drosophila silvestris and Drosophila heteroneura

Sister taxa endemic to the youngest island, Hawaii, provide some of the best examples of "neospecies" because they must have diverged less than 400,000 years ago. Most information is available for the species pair *D. silvestris* and *D. heteroneura*. These two sympatric species in the *planitibia* species subgroup show extremely high genetic similarity as judged by allozyme data (Sene and Carson, 1977; Craddock and Johnson, 1979),

DNA–DNA hybridization (Hunt and Carson, 1983), mtDNA restriction analysis (DeSalle *et al.*, 1986) and nucleotide sequence data (Rowan and Hunt, 1991; Kambysellis *et al.*, 1995), all of which confirm their recent divergence. Their isolation depends solely on a strong premating barrier because F_1, F_2, and backcross hybrids of both sexes are fully fertile (Craddock, 1974b; Ahearn and Val, 1975). As in many pairs of Hawaiian species, the premating isolation is asymmetrical, with the cross *silvestris* ♀ × *heteroneura* ♂ occurring more readily than the reciprocal (Craddock, 1974b; Ahearn and Val, 1975; Price and Boake, 1995). These broadly sympatric and interfertile species have remained distinct, despite occasional breakdown of the premating isolation; in the two populations where hybridization has been detected, natural hybrids comprise less than 2% of the population (Kaneshiro and Val, 1977; Carson *et al.*, 1989). Elsewhere behavioral isolation appears to be absolute, facilitating their continued divergence.

Given their extreme allozymic similarity and their genetic compatibility as evidenced by full hybrid fertility, just what are the critical genetic differences that separate this pair of neospecies? According to a variety of evidence (e.g., Craddock and Johnson, 1979; Hunt and Carson, 1983; DeSalle *et al.*, 1997), *D. silvestris* and *D. heteroneura* are sister species that derived from a single founder from the Maui complex. Hence they diverged on Hawaii from a common ancestral population, most likely at Hualalai. Their genetic divergence affected morphological, behavioral, and ecological characters, with the change in male head shape being the most striking (see the unusual hammer-shaped head of *D. heteroneura* in Fig. 7). The

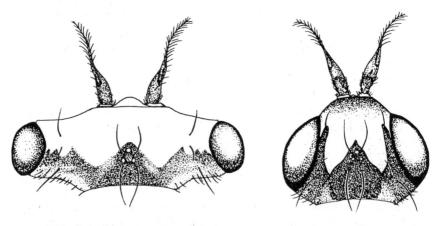

FIG. 7. Head morphology of adult males of the two sister taxa, *D. silvestris* and *D. heteroneura*, both from the island of Hawaii. Reproduced with permission of the University of Chicago Press from The American Naturalist, Vol. 111, K. Y. Kaneshiro and F. C. Val (1977).

genetic basis of these interspecific differences can be analyzed because hybrids are fertile. Accordingly, Val (1977) estimated that from 15 to 19 loci could account for the morphological differences in five characters: male head shape (about 10 loci) and head size, male face color, mesopleural pigmentation, and wing spotting. The polygenic system involves the epistatic interaction of one or more major X-linked genes with several autosomal loci with predominantly additive effects; these morphological differences show a Type II genetic architecture (one or a few major-effect loci), in contrast with the Type I architecture (many minor-effect loci) of ethological differences in male cricket songs (Shaw, 1996a). The morphological divergence between *D. silvestris* and *D. heteroneura* is associated with a striking divergence in male aggressive behavior (Spieth, 1981), but experimental data suggest that aggressive behavior is not directly involved in maintaining species distinction (Price and Boake, 1995).

Surprisingly, action patterns of male courtship behavior in these two species are essentially similar (Spieth, 1981), although there is a qualitative difference in the initial frontal display stage (Spieth, 1978; Watson, 1979), quantitative differences in the wing vibrations in the subsequent head-under-wing (HUW) phase, and frequency differences in their acoustic signals (Hoikkala *et al.*, 1994). As well, Price and Boake (1995) suggested that differences in the relative activity levels of the two species may be significant, with the strong behavioral reproductive isolation between the species largely a consequence of the earliest stages of intersexual interaction. There are as yet no estimates of the number of genes underlying these ethological differences, although it would appear that genetic divergence in courtship behavior is less than that in aggressive behavior.

Ecologically, *D. silvestris* prefers darker areas of the forest and extends to higher altitudes than *D. heteroneura*. Further, differences in the ultrastructure of the egg chorion (Kambysellis, 1974) are suggestive of a shift in oviposition site (Kambysellis, 1993), even though both species utilize some of the same breeding substrates. Overall, it can be inferred that a substantial number of genes currently distinguish these two recently diverged species. What is not clear is which genetic changes were primary and crucial to the speciation event, and which ones followed separation of the gene pools of these two closely related species.

Differentiation within D. silvestris

As an approach to this problem, it is instructive to examine the intraspecific differentiation within *D. silvestris* that may presage a future speciation event. Two morphotypes have been discovered that differ in a secondary sexual character of the male foreleg. Males from derived northern and eastern populations have an additional row of cilia on the tibia, a new trait acquired since the splitting of *D. silvestris* from *D. heteroneura* (Carson and Bryant, 1979). It is important to note that males vibrate their tibiae and tarsi against the dorsum of the female's abdomen during the HUW phase of courtship (Spieth, 1978). Thus this morphological change is integral to a behavioral change, and it can be inferred that the primary divergence entailed the coevolution of mating behavior and morphology under intense sexual selection. It should be noted, however, that cilia number may not be the direct target of sexual selection, but rather a pleiotropic effect of selection on some other character (Kaneshiro, 1989). The genetic basis of this ciliary difference between morphotypes is polygenic, with about 30% of the difference due to X-linked genes and the remainder due to genes on at least two autosomes (Carson and Lande, 1984).

Comparisons with closely related species in the *planitibia* subgroup, and mating asymmetries in reciprocal crosses between allopatric populations of *D. silvestris* (Kaneshiro and Kurihara, 1981), interpreted according to Kaneshiro's (1976b, 1980) hypothesis, indicate that the form with two rows of cilia is ancestral and the one with three rows is derived. The mating asymmetries help trace the geographic spread of *D. silvestris* from volcano to volcano on the island of Hawaii (Kaneshiro and Kurihara, 1981), with each successive colonization necessarily due to a founder event. Significantly, the proposed sequence is consistent with interpretations of the chromosomal inversion data (Craddock and Carson, 1989) and the mtDNA data (DeSalle and Templeton, 1988). In this case, it is clear that males of the derived form have acquired a novel morphological (and by inference, behavioral) element; strangely, this contradicts Kaneshiro's (1976b, 1980) central thesis that the basis of the mating asymmetry is loss or simplification of an element of courtship behavior in the derived males. Regardless, it appears that the initial differentiation in both the speciation of *D. silvestris* and *D. heteroneura* and the incipient speciation within *D. silvestris* involves the flies' sexual behavior and morphological correlates in male secondary sexual characters. Thus the primary reproductive barriers in the recent intraisland speciation of these Hawaiian *Drosophila* would appear to be ethological premating barriers; postzygotic isolation may or may not develop later depending on whether ecological shifts and other evolutionary changes take place within the separated gene pools.

Hawaiian Drosophilid Phylogeny

The broad speciation patterns within the Hawaiian *Drosophila* and specifically within the better studied picture-winged group can best be visualized by a phylogenetic analysis of the species relationships. The initial markers used to trace relationships were inversion differences in the polytene chromosomes (Carson, 1970, 1983, 1992). Chromosomal banding sequences have been most informative in identifying major groups and subgroups within the picture-wings, but because chromosomal arrangements are relatively conservative, as evidenced by the frequency of clusters of homosequential species, these data have not always been able to resolve the detailed relationships within subgroups. Nonetheless, the chromosomal inversion data demonstrate that the proliferation of picture-winged species has involved both intraisland and interisland speciation events, with a minimum of 45 interisland colonizations required to explain the relationships of the 106 species surveyed (Carson, 1992). Interisland founder events have thus been prominent in the evolution of the group; the majority of these have been from an older to a younger island but some back migrations have also been proposed (Carson 1983, 1992). It should be noted here that although useful for inferring relationships, fixation of inversion differences has been incidental to speciation among Hawaiian flies and such differences between related species contribute little if anything to reproductive isolation.

The molecular phylogenies that are currently available include fewer species, but because they are based on a larger number of informative characters, they can often provide greater resolution than the chromosomal phylogeny. Sequence data, however, are generally derived from a single individual per species, with the drawback that polymorphic versus fixed sequence differences are not resolved and are ignored in the analysis. Two of the questions that have been addressed using the molecular data concern the time of origin of the Hawaiian Drosophilidae and the number of original founders. All the data concur in suggesting an ancient origin predating the formation of the current high islands, but the interpretations vary. On the basis of a molecular clock for the alcohol dehydrogenase (*Adh*) gene and *Adh* nucleotide sequences of picture-winged and non-picture-winged representatives of the drosophiloids together with one representative of the Hawaiian scaptomyzoids, Thomas and Hunt (1991) inferred two separate introductions to Hawaii: one a little over 25 Mya leading to the scaptomyzoid lineage, and the second about 10 Mya initiating the drosophiloid lineage. The latter implies a founder event on the island of Necker, a low eroded island well to the northwest of Kauai that no longer supports forest or drosophilid populations.

The phylogeny derived from the mtDNA sequence data of DeSalle (1992) considered in the light of the fossil evidence suggests, however, a single colonization of the Hawaiian islands occurring at least 30 Mya. This would place the site of the original drosophilid founder event on a now submerged island northwest of Midway Island (see Fig. 1). This interpretation of a single ancestor is supported by an independent molecular phylogeny based on sequences of the nuclear gene *Yp1* (Kambysellis *et al.*, 1995). It is also in general agreement with the phylogenetic inference derived from immunological distances of a larval hemolymph protein, which suggests that the Hawaiian Drosophilidae originated from a single colonization about 42 Mya (Beverley and Wilson, 1985). Whatever the actual date and location of the original drosophilid founder, it is clear that the current fauna is but a fraction of all the endemic Hawaiian drosophilids that must have speciated in the Hawaiian archipelago; parenthetically, we can no longer trace any specific speciation event more ancient than 5.1 Mya, because all of the earlier fauna is now extinct.

Molecular analyses of the extant fauna support a pattern of sequential colonization from older to younger islands in each group and subgroup of flies, in general agreement with inferences from the chromosomal data but with some exceptions; specifically, the molecular phylogenies require fewer back migrations (DeSalle, 1995; Kambysellis *et al.*, 1995). Apart from this, the general congruence between molecular and nonmolecular phylogenies is remarkable.

Figure 8 shows a phylogeny of a sample of 44 Hawaiian species (including species from the more primitive white-tip-scutellum, *antopocerus,* modified-tarsi, and modified-mouthparts groups as well as picture-wings), based on nucleotide sequences of mtDNA and one of the vitellogenin or *Yolk protein* genes (Kambysellis and Craddock, 1997). The resolution provided by this phylogenetic tree allows us to examine the historical sequence of events in the ecological diversification in breeding substrates and correlated reproductive strategies. The ecological character reconstruction on the molecular tree suggests that the original adaptive radiation into diverse oviposition substrates occurred relatively rapidly, given the short branch lengths at the base of the tree, and quite early in the history of drosophilids in Hawaii (Kambysellis *et al.*, 1995). Data are not yet available for the scaptomyzoids, but among the Hawaiian drosophiloids the most primitive used fungi as a breeding niche, with subsequent invasions of decaying leaves, fruit, stems, bark, and tree flux niches (Fig. 8). Following each adaptive shift, adaptation to each breeding site was achieved via modifications of the female reproductive system including changes in ovarian structure and functioning (Kambysellis and Heed, 1971; Kambysellis *et al.*, 1995), changes in chorion ultrastructure (Kambysellis, 1993), and changes in ovipositor form (Craddock and Kambysellis, 1997). Clearly, a complex of many genes underlies the coordinated behavioral, morphological, and physiological changes in the adaptive divergence of these flies. Although ancient speciation events in the initial radiation of drosophilids in Hawaii seem to have been associated with ecological adaptive shifts, fewer host shifts have accompanied the more recent speciation events. The shift from use of tree

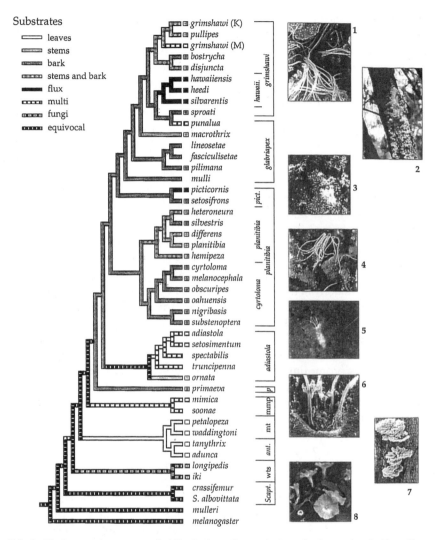

FIG. 8. Phylogenetic pattern of shifts in breeding substrate in the endemic Hawaiian *Drosophila*. Representative oviposition sites to the right show eggs in a tree flux (1, 2), soil flux (3), decaying bark (4), decaying leaf (5, 6), fungi (7), and flowers (8). MacClade (W. P. Maddison & Maddison, 1992) was used to map substrate use on a molecular phylogeny of 43 endemic Hawaiian *Drosophila* species and one Hawaiian *Scaptomyza* (*S. albovittata*), with two continental *Drosophila* (*D. melanogaster* and *D. mulleri*) as outgroups. The Hawaiian species groups and subgroups are as indicated to the right of the species names. The taxon indicated as *D. grimshawi* from Kauai (K) is now recognized as a sibling species of the Maui (M) *D. grimshawi* (see text, pp. 26–29). Reproduced from the article of Kambysellis and Craddock (1997) in *Molecular Evolution and Adaptive Radiation* edited by T. J. Givnish and K. J. Sytsma (ISBN 0-521-57329-7), with the permission of Cambridge University Press.

fluxes to soil fluxes in the sympatric species pair *D. silvarentis* and *D. heedi* on the island of Hawaii is a notable exception (Kambysellis and Craddock, 1997). Each of the various adaptive strategies has proved successful, a particular breeding niche and pattern of female reproductive traits being retained throughout a series of subsequent speciation events. The proliferation of species within the fungus-breeding, leaf-breeding, bark-breeding, and other ecological lineages must depend on other factors, most notably founder events and sexual selection as discussed more fully in the following section.

SPECIATION PROCESSES IN HAWAII

The preceding parade of examples of speciation in Hawaiian organisms is but a small sample of the many fascinating cases that have been studied. The characteristic pattern of adaptive radiation generating many ecologically diverse taxa from a single rare founder begs the central question: Why have species multiplied so readily in Hawaii? The answer is not a simple one, because there is a multiplicity of factors that affect the probability of a speciation event, and different factors play a greater or lesser role depending on the biology of the particular organism. Three of the most important evolutionary forces promoting speciation are natural selection, sexual selection, and random drift. Needless to say, these forces are not mutually exclusive, and their interaction may be more critical to the probability of a speciation event than the operation of any one factor alone.

The Role of Ecological Factors in Hawaiian Speciation

In both plant and animal groups, speciation in Hawaii is consistently associated with dramatic morphological and ecological differentiation. Many of the morphological changes are directly correlated with a particular ecological shift. This is clearly evident in the distinctive traits of cave-dwelling crickets and planthoppers, in the various growth habits of plants, and in the varied bill shapes of the honeycreepers, for example. The one apparent exception is the sword-tail crickets, a group in which sibling species abound. In each of the other examples considered, the ecological release afforded by movement into a different habitat or a new adaptive zone seems to have accelerated speciation rates and morphological differentiation. Colonization of open habitats in a newly formed island provides maximal ecological opportunity free of the competition that normally constrains

speciation events on continents. Much of the exuberance of insular speciation and of Hawaiian speciation in particular can be attributed to the heterogeneity of island environments and the access to new ecological niches
this affords. The general rule is that each shift into a new adaptive zone has
led to a burst of speciation. In many cases these shifts have occurred rapidly
and relatively early in the history of a group in Hawaii, as in the divergence
in breeding substrates in the drosophilids (Kambysellis *et al.*, 1995) and the
divergence in diet in the honeycreepers, these ecological shifts being pivotal
to the adaptive radiation of the group as a whole. The initial adaptive shifts
do not preclude additional shifts later in the evolutionary history of the
group as further novel environments are encountered. One example is the
shift from bark-breeding to flux-breeding within the *grimshawi* group of
Hawaiian *Drosophila* (Kambysellis and Craddock, 1997) that triggered a
new round of speciation events and evolution of a new clade, the *hawaiien-
sis* subgroup (Fig. 8). Likewise, speciation and adaptation have followed a
more recent shift from breeding in tree fluxes to breeding in soil fluxes
(Kambysellis and Craddock, 1997). Successful exploitation of these various
resources by Hawaiian *Drosophila* suggests that their adaptive radiation is
rooted in ecological processes, similarly to other cases where the ecological
interactions have been better studied (Schluter, 1996).

In Hawaii habitat shifts have been particularly critical to intraisland
speciation events, a point well demonstrated by the silversword, *Bidens*, and
Alsinoideae radiations. Expansion from more mesic low-altitude environments into more xeric high-altitude environments, for example, has invari-
ably triggered the evolution of new plant species adapted to the new
habitats in response to divergent natural selection. Although the genetic
change accompanying such shifts may initially be quite limited, the habitat
isolation is often sufficient to keep the gene pools in the two environments
separate, facilitating their further divergence.

Where the alternate habitats are not necessarily geographically separate (such as diverse breeding substrates within a forest in the case of
drosophilids, or food resources in the case of honeycreepers), speciation
may have been essentially sympatric. That substantial reproductive isolation can result from the evolution of habitat specialization via disruptive
selection on habitat preference has been clearly demonstrated experimentally (Rice, 1985) and via simulation (Rice, 1984). Rice and Hostert (1993)
summarized additional experimental evidence supporting the feasibility of
reproductive divergence under strong divergent selection despite gene flow.
Although considered a rare mode of speciation due to its restrictive
requirements (Templeton, 1981), ecological conditions in Hawaii may
have often favored sympatric speciation via microhabitat divergence or
host plant shifts (Asquith, 1995). Once some initial divergence had been

achieved, the continuing action of natural selection would further improve the adaptation of each incipient species to its particular ecological niche, leading to the remarkable adaptations observed in the current Hawaiian flora and fauna.

Although primarily an intraisland phenomenon, ecological shifts may also be involved in interisland speciation events. For example, the allopatric speciation of *D. grimshawi* on Maui Nui was accompanied by a shift from ovipositional specialist to ovipositional generalist, with a broadening of the breeding niche, an infrequent event among Hawaiian *Drosophila*. In other groups, ecological shifts occurring on one Hawaiian island may be replicated on another island, resulting in new and different species on each island. There is abundant evidence among plants for parallel adaptive shifts on different islands. Likewise, among animals parallel shifts have frequently been responsible for evolution of similar but distinct species on different islands; one of the best examples involves the independent evolution of cave crickets and cave-dwelling planthoppers on Maui Nui and Hawaii, resulting from parallel invasions of the subterranean habitat by surface-dwelling forms.

With their unique geology the Hawaiian islands provide an unusually broad array of habitats concentrated in small areas, and it is clear from the many spectacular radiations that ecological opportunity has been a major factor in the speciation of the Hawaiian flora and fauna. The driving force for the evolution of reproductive isolation in this instance is natural selection, operating pleiotropically via divergent selection between alternate habitats, either allopatrically or nonallopatrically (Rice, 1987; Rice and Hostert, 1993).

It is important to note that biotic factors may have been just as important as abiotic factors in determining evolutionary outcomes in many groups of Hawaiian organisms, and hence greater attention needs to be paid to the evolutionary role of species interactions. The composition of the community that a founder colonizes will affect its chance of successful establishment by determining the degree of ecological opportunity or competition, as well as the resulting patterns of coevolution with other community members. This is especially important in the case of herbivorous insects where the identity of the resident plants will certainly affect the outcome. The one example discussed of cospeciation between the *Nesosydne* planthoppers and their silversword hosts is likely to be a common pattern, given the interactions and interdependence of plants and insects.

The availability of a variety of plant resources, including the lobelioids, has clearly been important to the evolutionary divergence of the Hawaiian *Drosophila*. Morphological evolution of certain lobelioid species of *Cyanea* and *Rollandia*, on the other hand, may have been affected by the coexistence of browsing geese and ducks (Givnish *et al.*, 1994). Lobelioid evolution (and extinction) may have also been affected by evolution of their honeycreeper pollinators (Givnish *et al.*, 1995). This example linking specific plants, flies, and two groups of birds emphasizes that there is a complex web of ecological interactions in Hawaiian communities and that no species radiation should be considered in isolation; rather, coevolution is likely to be the norm, with reciprocal adaptation guided by natural selection an important component of the interactions among all elements of the flora and fauna.

The Role of Sexual Selection

Not all of the morphological differences distinguishing closely related Hawaiian species can be attributed to ecological shifts and the forces of natural selection. Among Hawaiian drosophilid species, for example, some of the most prominent morphological differences are secondary sexual characters of males, females of closely related species often being almost indistinguishable. This fact, the characteristic lek behavior of these flies, and their complex courtship routines suggest that sexual selection may be a major causal factor in their speciation (Spieth, 1968, 1974; Ringo, 1977; Carson, 1978, 1997; Kaneshiro and Boake, 1987). It should be noted, however, that Templeton (1979) has argued vigorously against this notion, although he does concede that sexual selection may contribute to the rapidity of speciation, but only in the context of a founder event. Sexual selection results from variance in mating success and is most often due to male–male competition and/or variation in female preferences for particular male traits. There is both laboratory and field evidence of intense sexual selection among males in a number of Hawaiian *Drosophila* species (Spiess and Carson, 1981; Carson, 1987; Shelly, 1987, 1988; Droney, 1992; Boake *et al.*, 1997), but the major factor responsible is less clear. Female choice has been widely emphasized (Kaneshiro, 1989), and most certainly operates because mating takes place at the leks, not at food sites (Spieth, 1974), and females only go to leks and mate when sexually receptive (Kambysellis and Craddock, 1991). Nonetheless, in *D. grimshawi* laboratory evidence is weak that female choice plays an active role in male mating success (Droney, 1992). In *D. heteroneura*, on the other hand, females apparently prefer males with broader heads, with strong directional selection on head width operating through mate choice (Boake *et al.*, 1997). As well, male head width correlates strongly with aggressive success, confirming that this male trait (Fig. 7) is sexually selected. Hawaiian *Drosophila* species vary, however, in the level of agonistic behavior between the resident male at a lek and intruders, as well as in other aspects of their lek biology (Shelly, 1987, 1988; Droney, 1992, 1994; Droney and Hock, 1998). Whereas most picture-winged species have solitary leks, in some species such as *D. grimshawi* several males routinely aggregate and interact at a lek but do not defend individual territories. These behavioral differences are undoubtedly associated with interspecific differences in the relative importance of male competition and female choice.

The contribution of sexual selection to speciation in the Hawaiian *Drosophila*, in the Hawaiian sword-tail crickets, and in the *Oliarus* planthoppers is difficult to disentangle from the effects of random drift and natural selection that must have accompanied speciation episodes. In each of these groups where ethological barriers effectively isolate sympatric (and allopatric) species, shifts in the mate recognition system seem to have been central to many speciation events. Male traits and female preferences (or, in the case of planthoppers, female traits and male preferences) cannot evolve independently, however, because coadaptation between the sexes is essential to successful reproduction; moreover, male and female traits are constrained by genetic correlations between the sexes (Lande, 1987). Most populations are assumed to contain ample genetically based behavioral variation within each sex (Kaneshiro, 1989), but a directional change in mating behavior is unlikely to occur without some perturbation. A founder event, for example, may temporarily destabilize the interactions between the sexes (see the following paragraph). Random genetic drift interacting with natural and sexual selection may then lead to rapid behavioral diversification as a new equilibrium is established. Quantitative genetic models (Lande, 1981) suggest that such behavioral shifts may lead to rapid speciation, and thus sexual selection has been posited as a driving force in the rapid speciation of Hawaiian *Drosophila* (Carson, 1978, 1997). The question is whether sexual selection, by itself, is sufficient to bring about speciation. If so, male traits that are sexually selected within a species should serve for species recognition and isolation between closely related sympatric species. Surprisingly, Boake *et al.* (1997) demonstrated that male head width in *D. heteroneura*, although sexually selected, is not a major contributor to behavioral isolation between the sympatric and sister taxa *D. heteroneura* and *D. silvestris*.

A specific model relating shifts in sexual behavior to the speciation process has been proposed by Kaneshiro (1989), largely to account for the mating asymmetries observed in reciprocal crosses among ancestral and derived species of Hawaiian *Drosophila* (Kaneshiro, 1976b). Evidence from laboratory mating tests and selection experiments in *D. silvestris* indicates that there is a range of mating types segregating within a population, with female discrimination varying from low to high, and male mating ability varying from poor to excellent, the two traits being strongly correlated. Following a founder event, there would be strong selection against highly discriminant females, with a consequent increase in the frequency of less discriminant females as long as population size remained small. Because of the genetic correlations between the sexes, males of reduced mating ability would accumulate, their lower success in mating presumably resulting from a decrease in the cumulative signal strength of their courtship stimuli. The latter is assumed to be the outcome of a simplification of their courtship or even a loss of specific courtship elements (Kaneshiro, 1976b, 1983). Once population size increased in the ensuing flush phase, differential selection between the sexes would be reinstituted with selection for higher mating ability in males and lower discrimination in females. These antagonistic forces would regenerate a wider array of mating types and drive the population to a new dynamic equilibrium, that in all probability differs from that in the ancestral population from which the founders were derived. As a result of shifts in the sexual selection process triggered by the founder event, the change in the sexual environment and in the mate recognition system between the ancestral and derived populations would engender a measure of sexual isolation between them. Continued operation of sexual selection and natural selection could then promote further divergence to complete reproductive isolation and full speciation. This model relies on the genetic and selective perturbation produced by a founder event, the corollary being that the sexual behavioral shift takes place allopatrically.

Founder events are also implicated in the ethological divergence among cave-adapted *Oliarus* planthoppers that has apparently occurred rapidly following their subterranean spread to new lava tubes (Hoch and Howarth, 1993). Here sexual selection must have operated on both males and females, with mutual tuning or coadaptation between the sexes essential, because in planthoppers each sex both transmits and receives acoustic signals (Den Hollander, 1995). In Hawaiian *Drosophila* and in Hawaiian crickets, only the male sex is assumed to be the target of sexual selection, although coadaptation between male signaling and female perception of those signals is clearly necessary to a functional system and to a significant behavioral shift in any derived population.

In the swordtail crickets where species proliferation has been largely intraisland, and indeed "intravolcano" or even "intravalley," founder events cannot necessarily be invoked as the causative factor for the ethological differences that distinguish species. Shifts into new behavioral and bioacoustic niches must, in some cases, have taken place sympatrically (Otte, 1989).

Sexual selection may have played a role in the divergence in cricket mate recognition systems, but thus far no specific models have been proposed to count for their speciation patterns.

e Role of Founder Events

Chance has played a significant role in the origin and diversification of the Hawaiian biota. Indeed Hawaiian speciation is centered around the theme of the founder principle (Carson, 1971). Rare founder events are implicit in the initial colonization of the remote Hawaiian archipelago, in the subsequent spread of living beings down the island chain, in their dispersion to newer volcanoes within an island, and even in the repeated recolonizations of volcanically devastated areas. All founder populations must be subject to random genetic drift at the outset because of their small size, but the loss of genetic variability may not be as severe as previously thought (Lande, 1980; Carson and Wisotzkey, 1989). The contention (Barton and Charlesworth, 1984) is not whether founder events have occurred, but whether the genetic bottlenecks and drift concomitant with founder events have directly induced species divergence and have been responsible for a unique mode of speciation, the so-called founder effect. Genetic drift is a central element in all models of founder effect speciation, but the several hypotheses vary in their emphases on the accompanying genetic revolution (Mayr, 1954), the genetic disorganization engendered by changing selection pressures associated with founder–flush–crash cycles (Carson, 1968, 1982), the perturbation in the genetic system leading to a genetic transilience and rapid shift to a new adaptive peak (Templeton, 1980, 1981), or the interactions between drift and sexual selection leading to ethological isolation (Kaneshiro, 1989). These alternate mechanisms are discussed at length in various reviews of founder effect speciation (Templeton, 1980, 1981; Carson and Templeton, 1984; Barton and Charlesworth, 1984; Barton, 1989; Spencer, 1995).

Founder events and rapid speciation are hallmarks of the speciation process in Hawaii but the causal link between them is by no means clear. It is naive to think that a founder event will automatically trigger speciation, and indeed there are enough cases known where dispersal to a new island has not resulted in speciation. Moreover, there is little experimental support for simple bottleneck-induced speciation despite extensive laboratory studies aimed at substantiating this mechanism (reviewed in Rice and Hostert, 1993; see also Galiana et al., 1993). Nonetheless, founder events seem to have been a necessary precondition for many of the speciation episodes in Hawaii, the founder event permitting the action of genetic drift, along with an altered regime of sexual and natural selection in the new locale. Because dispersal to a new island, a new volcano, or a new lava flow presents the founder with a relatively unexploited habitat with a community composition different from that in the source area, a change of environment with an attendant change in selective forces is an inevitable outcome of a founder event. Thus it is well nigh impossible to dissociate the effects of drift and selection on the founder population. Drift, a change in natural selection, and often a shift in sexual selection are all likely consequences of founder events, and realistically, the joint action of all of these forces must be considered part and parcel of founder effect speciation. To unravel the contribution of each to an individual speciation event or to Hawaiian speciation in general would seem to be problematic, and hence the actual mechanism of founder effect speciation remains controversial.

More recently, Slatkin (1996) has presented several plausible population genetic models of the founder–flush process that validate the likelihood of rapid genetic evolution following a founder event. Even peak shifts that could lead to rapid adaptation and speciation are quite likely, provided that the appropriate alleles are present in the newly founded population. The key to Slatkin's simple models is that genetic drift is weak in the flush phase of rapid growth, whereas natural selection is more effective under these conditions than in a population of constant size. Advantageous alleles with additive, recessive, and epistatic effects are shown to have a much greater chance of fixation during periods of rapid growth. The same conditions that facilitate the operation of natural selection should also make sexual selection more effective (Slatkin, 1996), so that ethological shifts generated by female preferences for novel male traits should also be more likely than if the founder event had not occurred.

Despite ongoing debates on the likelihood and actual genetic mechanism, founder effect speciation has clearly been a major mode of species proliferation in Hawaii, although globally rare as a general mechanism. In Hawaii, the geologic, volcanic, and ecological situation has provided the opportunity, and the successful colonizers have had the appropriate biological and genetic attributes (Templeton, 1980) to respond, adapt, and speciate in the insular situation. Certain groups such as the Hawaiian *Drosophila* have an ensemble of traits including high reproductive potential (Kambysellis and Heed, 1971), subdivided population structure (Craddock and Carson, 1989), and a complex sexually selected mate recognition system (Spieth, 1974; Kaneshiro, 1989) particularly conducive to rapid speciation following founder events. Over the past few million years, rare interisland founders and more numerous intraisland founders have established innumerable new populations, many of which underwent population crashes or extinction as a result of fortuitous volcanic factors. A minority survived and underwent the critical evolutionary shifts that led to novel species and the spectacular radiation that makes this group such an outstanding example of insular speciation.

SUMMARY

The Hawaiian Islands provide a showcase for the processes of evolutionary divergence and speciation, as demonstrated by the spectacular examples from plants, birds, and insects presented in this chapter. Rapid, explosive speciation has been driven by a combination of founder events, adaptive divergence to novel habitats, and in animals, behavioral divergence via shifts in sexual selection, that together have led to varying degrees of prezygotic and postzygotic isolation between ancestral and derived taxa. The fascinating biota of the Hawaiian Archipelago is unique, and the spectrum of biological variation and adaptation in Hawaiian organisms is a precious resource for evolutionary biologists. The knowledge we now have of the flora and fauna of these oceanic islands would have astounded Darwin, while confirming his prediction that their biology is "well worth examining" for insights into the complexities of the speciation process.

ACKNOWLEDGMENTS

I am indebted to Hampton L. Carson and D. Elmo Hardy who first introduced me to the Hawaiian biota, supporting my initial participation in the Hawaiian Drosophila Project with an NSF Postdoctoral Fellowship. I also acknowledge Hamp for his comments and suggestions on an early draft of this manuscript. Special thanks to my husband Michael Kambysellis for inspiration, insights into Hawaiian flies, and innumerable discussions over the years. To the many other colleagues who share an interest in Hawaiian evolutionary problems, my thanks for stimulating my thinking about speciation processes via interactions in the field, in the lab, and in the library.

REFERENCES

Ahearn, J. N., and Val, F. C., 1975, Fertile interspecific hybrids of two sympatric Hawaiian *Drosophila, Genetics* **80:**s9.

Amadon, D., 1950, The Hawaiian honeycreepers (Aves, Drepaniidae), *Bull. Amer. Mus. Nat. History* **95:**151–262.

Asche, M., 1997, A review of the systematics of Hawaiian planthoppers (Homoptera: Fulgoroidea), *Pacif. Sci.* **51:**366–376.

Asquith, A., 1995, Evolution of *Sarona* (Heteroptera, Miridae), Speciation on geographic and ecological islands, in: *Hawaiian Biogeography: Evolution on a Hot Spot Archipelago* (W. L. Wagner and V. A. Funk, Eds.), pp. 90–120, Smithsonian Institution Press, Washington and London.

Baldwin, B. G., 1997, Adaptive radiation of the Hawaiian silversword alliance: Congruence and conflict of phylogenetic evidence from molecular and non-molecular investigations, in: *Molecular Evolution and Adaptive Radiation* (T. J. Givnish and K. J. Sytsma, Eds.), pp. 103–128, Cambridge Univ. Press, Cambridge.

Baldwin, B. G., and Robichaux, R. H., 1995, Historical biogeography and ecology of the Hawaiian silversword alliance (Asteraceae). New molecular phylogenetic perspectives, in: *Hawaiian Biogeography: Evolution on a Hot Spot Archipelago* (W. L. Wagner and V. A. Funk, Eds.), pp. 259–287, Smithsonian Institution Press, Washington and London.

Baldwin, B. G., and Sanderson, M. J., 1998, Age and rate of diversification of the Hawaiian silversword alliance (Compositae) *Proc. Natl. Acad. Sci. USA* **95:**9402–9406.

Baldwin, B. G., Kyhos, D. W., and Dvorak, J., 1990, Chloroplast DNA evolution and adaptive radiation in the Hawaiian silversword alliance (Asteraceae–Madiinae), *Ann. Miss. Bot. Gard.* **77:**96–109.

Baldwin, B. G., Kyhos, D. W., Dvorak, J., and Carr, G. D., 1991, Chloroplast DNA evidence for a North American origin of the Hawaiian silversword alliance (Asteraceae), *Proc. Natl. Acad. Sci. USA* **88:**1840–1843.

Barton, N. H., 1989, Founder effect speciation, in: *Speciation and its Consequences* (D. Otte and J. A. Endler, Eds.), pp. 229–256, Sinauer Associates, Inc., Sunderland, MA.

Barton, N. H., and Charlesworth, B., 1984, Genetic revolutions, founder effects, and speciation, *Annu. Rev. Ecol. Syst.* **15:**133–164.

Beverley, S. M., and Wilson, A. C., 1985, Ancient origin for Hawaiian Drosophilinae inferred from protein comparisons, *Proc. Natl. Acad. Sci. USA* **82:**4753–4757.

Boake, C. R. B., DeAngelis, M. P., and Andreadis, D. K., 1997, Is sexual selection and species recognition a continuum? Mating behavior of the stalk-eyed fly *Drosophila heteroneura*, *Proc. Natl. Acad. Sci. USA* **94:**12442–12445.

Carlquist, S., 1970, *Hawaii: A Natural History*, The Natural History Press, Garden City, New York.

Carlquist, S., 1982, The first arrivals, *Nat. Hist.* **91(12):**20–30.

Carr, G. D., 1985, Monograph of the Hawaiian Madiinae (Asteraceae): *Argyroxiphium, Dubautia*, and *Wilkesia, Allertonia* **4:**1–123.

Carr, G. D., 1987, Beggar's ticks and tarweeds: Masters of adaptive radiation, *Trends in Ecol. & Evol.* **2:**192–195.

Carr, G. D., and Kyhos, D. W., 1986, Adaptive radiation in the Hawaiian silversword alliance (Compositae-Madiinae). II. Cytogenetics of artificial and natural hybrids, *Evolution* **40:**959–976.

Carr, G. D., Powell, E. A., and Kyhos, D. W., 1986, Self-incompatibility in the Hawaiian Madiinae (Compositae): An exception to Baker's rule, *Evolution* **40:**430–434.

Carr, G. D., Robichaux, R. H., Witter, M. S., and Kyhos, D. W., 1989, Adaptive radiation of the Hawaiian silversword alliance (Compositae–Madiinae): A comparison with Hawaiian picture-winged *Drosophila*, in: *Genetics, Speciation and the Founder Principle* (L. V. Giddings, K. Y. Kaneshiro and W. W. Anderson, Eds.), pp. 79–97, Oxford University Press, New York.

Carson, H. L., 1968, The population flush and its genetic consequences, in: *Population*

Biology and Evolution (R. C. Lewontin, Ed.), pp. 123–137, Syracuse Univ. Press, New York.

Carson, H. L., 1970, Chromosome tracers of the origin of species, *Science* **168:**1414–1418.

Carson, H. L., 1971, Speciation and the founder principle, *Stadler Genet. Symp.* **3:**51–70.

Carson, H. L., 1978, Speciation and sexual selection in Hawaiian *Drosophila*, in: *Ecological Genetics: The Interface* (P. F. Brussard, Ed.), pp. 93–107, Springer-Verlag, New York.

Carson, H. L., 1982, Speciation as a major reorganization of polygenic balances, in: *Mechanisms of Speciation* (C. Barigozzi, Ed.), pp. 411–433, Liss, New York.

Carson, H. L., 1983, Chromosomal sequences and interisland colonizations in Hawaiian *Drosophila, Genetics* **103:**465–482.

Carson, H. L., 1987, High fitness of heterokaryotypic individuals segregating naturally within a long-standing laboratory population of *Drosophila silvestris, Genetics* **116:**415–422.

Carson, H. L., 1992, Inversions in Hawaiian *Drosophila*, in: *Drosophila Inversion Polymorphism* (C. B. Krimbas and J. R. Powell, Eds.), pp. 407–439, CRC Press, Boca Raton.

Carson, H. L., 1997, Sexual selection: A driver of genetic change in Hawaiian *Drosophila, J. Heredity* **88:**343–352.

Carson, H. L., and Bryant, P. J., 1979, Change in a secondary sexual character as evidence of incipient speciation in *Drosophila silvestris, Proc. Natl. Acad. Sci. USA* **76:**1929–1932.

Carson, H. L., and Clague, D. A., 1995, Geology and biogeography of the Hawaiian Islands, in: *Hawaiian Biogeography: Evolution on a Hot Spot Archipelago* (W. L. Wagner and V. A. Funk, Eds.), pp. 14–29, Smithsonian Institution Press, Washington and London.

Carson, H. L., and Kaneshiro, K. Y., 1976, *Drosophila* of Hawaii: Systematics and ecological genetics, *Annu. Rev. Ecol. Syst.* **7:**311–345.

Carson, H. L., and Lande, R., 1984, Inheritance of a secondary sexual character in *Drosophila silvestris, Proc. Natl. Acad. Sci. USA* **81:**6904–6907.

Carson, H. L., and Templeton, A. R., 1984, Genetic revolutions in relation to speciation phenomena: The founding of new populations, *Annu. Rev. Ecol. Syst.* **15:**97–131.

Carson, H. L., and Wisotzkey, R. G., 1989, Increase in genetic variance following a population bottleneck, *Amer. Nat.* **134:**668–673.

Carson, H. L., Hardy, D. E., Spieth, H. T., and Stone, W. S., 1970, The evolutionary biology of the Hawaiian Drosophilidae, in: *Essays in Evolution and Genetics in Honor of Theodosius Dobzhansky* (M. K. Hecht and W. C. Steere, Eds.), pp. 437–543, Appleton-Century-Crofts, New York.

Carson, H. L., Kaneshiro, K. Y., and Val, F. C., 1989, Natural hybridization between the sympatric Hawaiian species *Drosophila silvestris* and *Drosophila heteroneura, Evolution* **43:**190–203.

Carson, H. L., Lockwood, J. P., and Craddock, E. M., 1990, Extinction and recolonization of local populations on a growing shield volcano, *Proc. Natl. Acad. Sci. USA* **87:**7055–7057.

Clague, D. A., and Dalrymple, G. B., 1987, The Hawaiian-Emperor volcanic chain. Part I. Geologic evolution, in: *Volcanism in Hawaii* (R. W. Decker, T. L. Wright and P. H. Stauffer, Eds.), USGS Professional Paper 1350, US Gov. Printing Office.

Claridge, M. F., 1985, Acoustic signals in the Homoptera: Behavior, taxonomy, and evolution, *Annu. Rev. Entomol.* **30:**297–317.

Cowie, R. H., 1995, Variation in species diversity and shell shape in Hawaiian land snails: *In situ* speciation and ecological relationships, *Evolution* **49:**1191–1202.

Craddock, E. M., 1974a, Degrees of reproductive isolation between closely related species of Hawaiian *Drosophila*, in: *Genetic Mechanisms of Speciation in Insects* (M. J. D. White, Ed.), pp. 111–139, Australia and New Zealand Book Co., Sydney.

Craddock, E. M., 1974b, Reproductive relationships between homosequential species of Hawaiian *Drosophila, Evolution* **28:**593–606.

Craddock, E. M., and Carson, H. L., 1989, Chromosomal inversion patterning and population differentiation in a young insular species *Drosophila silvestris*, *Proc. Natl. Acad. Sci. USA* **86:**4798–4802.

Craddock, E. M., and Dominey, W., 1998, Adult age and breeding structure of a Hawaiian *Drosophila silvestris* (Diptera: Drosophilidae) population assessed via female reproductive status, *Pacif. Sci.* **52:**197–209.

Craddock, E. M., and Johnson, W. E., 1979, Genetic variation in Hawaiian *Drosophila*. V. Chromosomal and allozymic diversity in *Drosophila silvestris* and its homosequential species, *Evolution* **33:**137–155.

Craddock, E. M., and Kambysellis, M. P., 1990, Vitellogenin protein diversity in the Hawaiian *Drosophila*, *Biochem. Genet.* **28:**415–432.

Craddock, E. M., and Kambysellis, M. P., 1997, Adaptive radiation in the Hawaiian *Drosophila* (Diptera: Drosophilidae): Ecological and reproductive character analyses, *Pacif. Sci.* **51:**475–489.

Darwin, C., 1835, Excerpt from letter to Nature, Sep. 7, 1835, cited in Barlow, N. (1946) *Charles Darwin and the Voyage of the Beagle*, Philosophical Library, New York.

Den Hollander, J., 1995, Acoustic signals as specific-mate recognition signals in leafhoppers (Cicadellidae) and planthoppers (Delphacidae) (Homoptera: Auchenorrhyncha), in: *Speciation and the Recognition Concept* (D. M. Lambert and H. G. Spencer, Eds.), pp. 440–463, The Johns Hopkins Univ. Press, Baltimore.

DeSalle, R., 1992, The origin and possible time of divergence of the Hawaiian Drosophilidae: Evidence from DNA sequences, *Mol. Biol. Evol.* **9:**905–916.

DeSalle, R., 1995, Molecular approaches to biogeographic analysis of Hawaiian Drosophilidae, in: *Hawaiian Biogeography: Evolution on a Hot Spot Archipelago* (W. L. Wagner and V. A. Funk, Eds.), pp. 72–89, Smithsonian Institution Press, Washington and London.

DeSalle, R., and Templeton, A. R., 1988, Founder effects and the rate of mitochondrial DNA evolution in Hawaiian *Drosophila*, *Evolution* **42:**1076–1084.

DeSalle R., Giddings, L. V., and Kaneshiro, K. Y., 1986, Mitochondrial DNA variability in natural populations of Hawaiian *Drosophila*. II. Genetic and phylogenetic relationships of *D. silvestris* and *D. heteroneura*, *Heredity* **56:**87–92.

DeSalle, R., Brower, A. V. Z., Baker, R., and Remsen, J., 1997, A hierarchical view of the Hawaiian Drosophilidae (Diptera), *Pacif. Sci.* **51:**462–474.

Dickinson, W. J., 1980, Tissue specificity of enzyme expression regulated by diffusible factors: Evidence in *Drosophila* hybrids, *Science* **207:**995–997.

Droney, D. C., 1992, Sexual selection in a lekking Hawaiian *Drosophila*: The roles of male competition and female choice in male mating success, *Anim. Behav.* **44:**1007–1020.

Droney, D. C., 1994, Tests of hypotheses for lek formation in a Hawaiian *Drosophila*, *Anim. Behav.* **47:**351–361.

Droney, D. C., and Hock, M. B., 1998, Male sexual signals and female choice in *Drosophila grimshawi* (Diptera: Drosophilidae), *J. Insect Behav.* **11:**59–71.

Fleischer, R. C., McIntosh, C. E., and Tarr, C. L., 1998, Evolution on a volcanic conveyer belt: Using phylogeographic reconstructions and K–Ar based ages of the Hawaiian Islands to estimate molecular evolutionary rates, *Molec. Ecol.* **7:**533–545.

Freed, L. A., Conant, S., and Fleischer, R. C., 1987, Evolutionary ecology and radiation of Hawaiian passerine birds, *Trends Ecol. & Evol.* **2(7):**196–203.

Funk, V. A., and Wagner, W. L., 1995, Biogeographic patterns in the Hawaiian Islands, in: *Hawaiian Biogeography: Evolution on a Hot Spot Archipelago* (W. L. Wagner and V. A. Funk, Eds.), pp. 379–419, Smithsonian Institution Press, Washington and London.

Galiana, A., Moya, A., and Ayala, F. J., 1993, Founder–flush speciation in *Drosophila pseudoobscura*: A large-scale experiment, *Evolution* **47:**432–444.

Ganders, F. R., 1989, Adaptive radiation in Hawaiian *Bidens*, in: *Genetics, Speciation, and the Founder Principle* (L. V. Giddings, K. Y. Kaneshiro and W. W. Anderson, Eds.), pp. 99–112, Oxford University Press, New York.

Givnish, T. J., 1997, Adaptive radiation and molecular systematics: Issues and approaches, in: *Molecular Evolution and Adaptive Radiation* (T. J. Givnish and K. J. Sytsma, Eds.), pp. 1–54, Cambridge Univ. Press, Cambridge.

Givnish, T. J., and Sytsma, K. J., Eds., 1997, *Molecular Evolution and Adaptive Radiation*, Cambridge Univ. Press, Cambridge.

Givnish, T. J., Sytsma, K. J., Smith, J. F., and Hahn, W. J., 1994, Thorn-like prickles and hetero-phylly in *Cyanea*: adaptations to extinct avian browsers on Hawaii? *Proc. Natl. Acad. Sci. USA* **91**:2810–2814.

Givnish, T. J., Sytsma, K. J., Smith, J. F., and Hahn, W. J., 1995, Molecular evolution, adaptive radiation, and geographic speciation in *Cyanea* (Campanulaceae, Lobelioideae), in: *Hawaiian Biogeography. Evolution on a Hot Spot Archipelago* (W. L. Wagner and V. A. Funk, Eds.), pp. 288–337, Smithsonian Institution Press, Washington.

Givnish, T. J., Knox, E., Patterson, T. B., Haperman, J. R., Palmer, J. D., and Sytsma, K. J., 1996, The Hawaiian lobelioids are monophyletic and underwent a rapid initial radiation roughly 15 million years ago, *Am. J. Bot.* **83** (Suppl.):159.

Grant, P. R., 1981, Speciation and the adaptive radiation of Darwin's finches, *Amer. Sci.* **69**:653–663.

Hardy, D. E., 1965, *Insects of Hawaii, Vol. 12, Diptera: Cyclorrhapha II. Series Schizophora, Section Acalypterae I. Family Drosophilidae*, Univ. Hawaii Press, Honolulu, Hawaii.

Hardy, D. E., and Kaneshiro, K. Y., 1981, Drosophilidae of Pacific Oceania, in: *The Genetics and Biology of Drosophila*, Vol. 3a. (M. Ashburner, H. L. Carson and J. N. Thompson, Eds.), pp. 309–348, Academic Press, New York.

Hatzopoulos, P., Craddock, E. M., and Kambysellis, M. P., 1989, DNA length variants contiguous to the 3′ end of a vitellogenin gene in *Drosophila grimshawi* laboratory stocks from different Hawaiian islands, *Biochem. Genet.* **27**:367–377.

Heed, W. B., 1968, Ecology of the Hawaiian Drosophilidae, *Univ. Texas Publ.* **6818**:387–419.

Heed, W. B., 1971, Host plant specificity and speciation in Hawaiian *Drosophila*, *Taxon* **20**:115–121.

Helenurm, K., and Ganders, F. R., 1985, Adaptive radiation and genetic differentiation in Hawaiian *Bidens*, *Evolution* **39(4)**:753–765.

Ho, K.-F., Craddock, E. M., Piano, F., and Kambysellis, M. P., 1996, Phylogenetic analysis of DNA length mutations in a repetitive region of the Hawaiian *Drosophila* Yolk protein gene *Yp2*, *J. Mol. Evol.* **43**:116–124.

Hoch, H., and Howarth, F. G., 1993, Evolutionary dynamics of behavioral divergence among populations of the Hawaiian cave-dwelling planthopper *Oliarus polyphemus* (Homoptera: Fulgoroidea: Cixiidae), *Pacif. Sci.* **47**:303–318.

Hoikkala, A., Kaneshiro, K. Y., and Hoy, R. R., 1994, Courtship songs of the picture-winged *Drosophila planitibia* subgroup species, *Anim. Behav.* **47**:1363–1374.

Howarth, F. G., 1972, Cavernicoles in lava tubes on the Island of Hawaii, *Science* **175**:325–326.

Howarth, F. G., 1983, Ecology of cave arthropods, *Annu. Rev. Entomol.* **28**:365–389.

Howarth, F. G., 1993, High-stress subterranean habitats and evolutionary change in cave-inhabiting arthropods, *Amer. Nat.* **142**:S65–S77.

Howarth, F. G., and Mull, W. P., 1992, *Hawaiian Insects and Their Kin*, University of Hawaii Press, Honolulu.

Howarth, F. G., Hoch, H., and Asche, M., 1990, Duets in darkness: Species-specific substrate-borne vibrations produced by cave-adapted cixiid planthoppers in Hawaii (Homoptera: Fulgoroidea), *Mem. Biospeol.* **17:**77–80.

Hunt, J. A., and Carson, H. L., 1983, Evolutionary relationships of four species of Hawaiian *Drosophila* as measured by DNA reassociation, *Genetics* **104:**353–364.

Johnson, N. K., Marten, J. A., and Ralph, C. J., 1989, Genetic evidence for the origin and relationships of Hawaiian honeycreepers (Aves: Fringillidae), *The Condor* **91:**379–396.

Kambysellis, M. P., 1974, Ultrastructure of the chorion in very closely related *Drosophila* species endemic to Hawaii, *Syst. Zool.* **23:**507–512.

Kambysellis, M. P., 1993, Ultrastructural diversity in the egg chorion of Hawaiian *Drosophila* and *Scaptomyza*: Ecological and phylogenetic considerations, *Int. J. Insect Morphol. & Embryol.* **22:**417–446.

Kambysellis, M. P., and Craddock, E. M., 1991, Insemination patterns in Hawaiian *Drosophila* species (Diptera: Drosophilidae) correlate with ovarian development, *J. Insect Behavior* **4:**83–100.

Kambysellis, M. P., and Craddock, E. M., 1997, Ecological and reproductive shifts in the diversification of the endemic Hawaiian *Drosophila*, in: *Molecular Evolution and Adaptive Radiation* (T. J. Givnish and K. J. Sytsma, Eds.), pp. 475–509, Cambridge Univ. Press, Cambridge.

Kambysellis, M. P., and Heed, W. B., 1971, Studies of oogenesis in natural populations of Drosophilidae. I. Relation of ovarian development and ecological habitats of the Hawaiian species, *Amer. Nat.* **105:**31–49.

Kambysellis, M. P., Hatzopoulos, P., and Craddock, E. M., 1989, The temporal pattern of vitellogenin synthesis in *Drosophila grimshawi*, *J. Exp. Zool.* **251:**339–348.

Kambysellis, M. P., Hatzopoulos, P., Seo, E. W., and Craddock, E. M., 1986, Noncoordinate synthesis of the vitellogenin proteins in tissues of *Drosophila grimshawi*, *Devel. Genet.* **7:**81–97.

Kambysellis, M. P., Ho, K.-F., Craddock, E. M., Piano, F., Parisi, M., and Cohen, J., 1995, Pattern of ecological shifts in the diversification of Hawaiian *Drosophila* inferred from a molecular phylogeny, *Current Biology* **5:**1129–1139.

Kaneshiro, K. Y., 1976a, A revision of generic concepts in the biosystematics of Hawaiian Drosophilidae, *Proc. Haw. Ent. Soc.* **22:**255–278.

Kaneshiro, K. Y., 1976b, Ethological isolation and phylogeny in the *planitibia* subgroup of Hawaiian *Drosophila*, *Evolution* **30:**740–745.

Kaneshiro, K. Y., 1980, Sexual isolation, speciation and the direction of evolution, *Evolution* **34:**437–444.

Kaneshiro, K. Y., 1983, Sexual selection and direction of evolution in the biosystematics of Hawaiian Drosophilidae, *Annu. Rev. Entomol.* **28:**161–178.

Kaneshiro, K. Y., 1989, The dynamics of sexual selection and founder effects in species formation, in: *Genetics, Speciation and the Founder Principle* (L. V. Giddings, K. Y. Kaneshiro and W. W. Anderson, Eds.), pp. 279–296, Oxford University Press, New York.

Kaneshiro, K. Y., and Boake, C. R. B., 1987, Sexual selection and speciation: Issues raised by Hawaiian *Drosophila*, *Trends Ecol. & Evolution* **2:**207–212.

Kaneshiro, K. Y., and Kambysellis, M. P., 1999, Description of a new allopatric sibling species of the Hawaiian picture-winged *Drosophila*, *Pacif. Sci.* **53:**208–213.

Kaneshiro, K. Y., and Kurihara, J. S., 1981, Sequential differentiation of sexual behavior in populations of *Drosophila silvestris*, *Pacif. Sci.* **35:**177–183.

Kaneshiro, K. Y., and Val, F. C., 1977, Natural hybridization between a sympatric pair of Hawaiian *Drosophila*, *Amer. Nat.* **111:**897–902.

Lack, D., 1947, *Darwin's Finches*, Cambridge Univ. Press, Cambridge.

Lande, R., 1980, Genetic variation and phenotypic evolution during allopatric speciation, *Amer. Nat.* **116**:463–479.

Lande, R., 1981, Models of speciation by sexual selection on polygenic traits, *Proc. Natl. Acad. Sci. USA* **78**:3721–3725.

Lande, R., 1987, Genetic correlations between the sexes in the evolution of sexual dimorphism and mating preferences, in: *Sexual Selection: Testing The Alternatives* (J. W. Bradbury and M. B. Andersson, Eds.), pp. 83–94, John Wiley & Sons, Chichester.

Lockwood, J. P., and Lipman, P. W., 1987, Holocene eruptive history of Mauna Loa volcano, in: *U.S. Geological Survey Professional Paper 1350*, pp. 509–535.

Macdonald, G. A., and Abbott, A. T., 1970, *Volcanoes in the Sea*, Univ. of Hawaii Press, Honolulu.

Maddison, W. P., and Maddison, D. R., 1992, *MacClade, Analysis of Phylogeny and Character Evolution, Version 3*, Sinauer Associates Inc., Sunderland, MA.

Martinez-Cruzado, J. C., Swimmer, C., Fenerjian, M. G., and Kafatos, F. C., 1988, Evolution of the autosomal chorion locus in *Drosophila*. I. General organization of the locus and sequence comparisons of genes *s15* and *s19* in evolutionarily distant species, *Genetics* **119**:663–677.

Mayr, E., 1954, Change of genetic environment and evolution, in: *Evolution as a Process* (J. Huxley, A. C. Hardy and E. B. Ford, Eds.), pp. 157–180, Allen & Unwin, London.

Miller, S. E., and Eldredge, L. G., 1996, Numbers of Hawaiian species: Supplement 1, *Bishop Mus. Occas. Pap.* **45**:8–17.

Montgomery, S. L., 1975, Comparative breeding site ecology and the adaptive radiation of picture-winged *Drosophila*, *Proc. Hawaii Entomol. Soc.* **22**:65–102.

Nei, M., Maruyama, T., and Chakraborty, R., 1975, The bottleneck effect and genetic variability in populations, *Evolution* **29**:1–10.

Nishida, G., Ed., 1994, Hawaiian terrestrial arthropod checklist, *Hawai'i Biological Survey*, Bishop Museum Press, Honolulu.

Nishida, G. M., 1997, Hawaiian terrestrial arthropod checklist. Third edition, *Bishop Mus. Tech. Rep.* **12**, iv+263 p. [http://www.bishop.hawaii.org/bishop/HBS/arthrosearch.html]

Normark, W. R., Clague, D. A., and Moore, J. G., 1982, The next island, *Nat. Hist.* **91**:68–71.

Ohta, A. T., 1978, Ethological isolation and phylogeny in the *grimshawi* species complex of Hawaiian *Drosophila*, *Evolution* **32**:485–492.

Ohta, A. T., 1980, Coadaptive gene complexes in incipient species of Hawaiian *Drosophila*, *Amer. Nat.* **115**:121–132.

Ohta, A. T., 1989, Coadaptive changes in speciation via the founder principle in the *grimshawi* species complex of Hawaiian *Drosophila*, in: *Genetics, Speciation and the Founder Principle* (L. V. Giddings, K. Y. Kaneshiro and W. W. Anderson, Eds.), pp. 315–328, Oxford University Press, New York.

Olson, S. L., and James, H. F., 1982, Fossil birds from the Hawaiian Islands: Evidence for wholesale extinction by man before Western contact, *Science* **217**:633–635.

Otte, D., 1989, Speciation in Hawaiian crickets, in: *Speciation and Its Consequences* (D. Otte and J. A. Endler, Eds.), pp. 482–526, Sinauer Associates, Inc., Sunderland, MA.

Page, R. D. M., 1995, Parallel phylogenies: Reconstructing the history of host–parasite assemblages, *Cladistics* **10**:155–173.

Piano, F., Craddock, E. M., and Kambysellis, M. P., 1997, Phylogeny of the island populations of the Hawaiian *Drosophila grimshawi* complex: Evidence from combined data, *Mol. Phylog. Evol.* **7**:173–184.

Price, D. K., and Boake, C. R. B., 1995, Behavioral reproductive isolation in *Drosophila silvestris*, *D. heteroneura*, and their F$_1$ hybrids, *J. Insect Behav.* **8:**595–616.

Raikow, R. J., 1976, The origin and evolution of the Hawaiian honeycreepers (Drepanididae), *The Living Bird* **15:**95–117.

Rice, W. R., 1984, Disruptive selection on habitat preference and the evolution of reproductive isolation: A simulation study, *Evolution* **38:**1251–1260.

Rice, W. R., 1985, Disruptive selection on habitat preference and the evolution of reproductive isolation: An exploratory experiment, *Evolution* **39:**645–656.

Rice, W. R., 1987, Speciation via habitat specialization: The evolution of reproductive isolation as a correlated character, *Evolutionary Ecology* **1:**301–314.

Rice, W. R., and Hostert, E. E., 1993, Laboratory experiments on speciation: What have we learned in forty years? *Evolution* **47:**1637–1653.

Ringo, J. M., 1976, A communal display in Hawaiian *Drosophila* (Diptera: Drosophilidae), *Ann. Entomol. Soc. Am.* **69:**209–214.

Ringo, J. M., 1977, Why 300 species of Hawaiian *Drosophila*? The sexual selection hypothesis, *Evolution* **31:**694–696.

Ringo, J. M., and Hodosh, R. J., 1978, A multivariate analysis of behavioral divergence among closely related species of endemic Hawaiian *Drosophila*, *Evolution* **33:**389–397.

Roderick, G. K., 1997, Herbivorous insects and the Hawaiian silversword alliance: Coevolution or cospeciation? *Pacif. Sci.* **51:**440–449.

Roderick, G. K., and Metz, E. C., 1997, Biodiversity of planthoppers (Hemiptera: Delphacidae) on the Hawaiian silversword alliance: Effects of host plant history and hybridization, *Mem. Mus. Vic.* **56:**393–399.

Rowan, R. G., and Hunt, J. A., 1991, Rates of DNA change and phylogeny from the DNA sequences of the alcohol dehydrogenase gene for five closely related species of Hawaiian *Drosophila*, *Mol. Biol. Evol.* **8:**49–70.

Sakai, A. K., Wagner, W. L., Ferguson, D. M., and Herbst, D. R., 1995, Origins of dioecy in the Hawaiian flora, *Ecology* **76:**2517–2529.

Sakai, A. K, Weller, S. G., Wagner, W. L., Soltis, P. S., and Soltis, D. E., 1997, Phylogenetic perspectives on the evolution of dioecy: Adaptive radiation in the endemic Hawaiian genera *Schiedea* and *Alsinidendron* (Caryophyllaceae: Alsinoideae), in: *Molecular Evolution and Adaptive Radiation* (T. J. Givnish and K. J. Sytsma, Eds.), pp. 455–473, Cambridge Univ. Press, Cambridge.

Schluter, D., 1996, Ecological causes of adaptive radiation, *Amer. Nat.* **148:**S40–S64.

Sene, F. M., and Carson, H. L., 1977, Genetic variation in Hawaiian *Drosophila* IV. Allozymic similarity between *D. silvestris* and *D. heteroneura* from the island of Hawaii, *Genetics* **86:**187–198.

Shaw, K. L., 1995, Biogeographic patterns of two independent Hawaiian cricket radiations (*Laupala* and *Prognathogryllus*), in: *Hawaiian Biogeography: Evolution on a Hot Spot Archipelago* (W. L. Wagner and V. A. Funk, Eds.), pp. 39–56, Smithsonian Institution Press, Washington D.C.

Shaw, K. L., 1996a, Polygenic inheritance of a behavioral phenotype: Interspecific genetics of song in the Hawaiian cricket genus *Laupala*, *Evolution* **50:**256–266.

Shaw, K. L., 1996b, Sequential radiations and patterns of speciation in the Hawaiian cricket genus *Laupala* inferred from DNA sequences, *Evolution* **50:**237–255.

Shelly, T. E., 1987, Lek behaviour of a Hawaiian *Drosophila*: Male spacing, aggression and female visitation, *Anim. Behav.* **35:**1394–1404.

Shelly, T. E., 1988, Lek behaviour of *Drosophila cnecopleura* in Hawaii, *Ecolog. Entomol.* **13:**51–55.

Simon, C., 1987, Hawaiian evolutionary biology: An introduction, *Trends Ecol. & Evolution* **2(7):**175–178.

Slatkin, M., 1996, In defense of founder–flush theories of speciation, *Amer. Nat.* **147:**493–505.

Soltis, P. S., Soltis, D. E., Weller, S. G., Sakai, A. K., and Wagner, W. L., 1996, Molecular phylogenetic analysis of the Hawaiian endemics *Schiedea* and *Alsinidendron* (Caryophyllaceae), *Syst. Bot.* **21:**365–379.

Spencer, H. G., 1995, Models of speciation by founder effect: A review, in: *Speciation and the Recognition Concept* (D. M. Lambert and H. G. Spencer, Eds.), pp. 141–156, The Johns Hopkins Univ. Press, Baltimore & London.

Spiess, E. B., and Carson, H. L., 1981, Sexual selection in *Drosophila silvestris* of Hawaii. *Proc. Natl. Acad. Sci. USA* **78:**3088–3092.

Spieth, H. T., 1966, Courtship behavior of endemic Hawaiian *Drosophila*, *Univ. Texas Publ.* **6615:**245–313.

Spieth, H. T., 1968, Evolutionary implications of sexual behavior in *Drosophila*, *Evolutionary Biology*, Vol. 2 (Th. Dobzhansky, M. K. Hecht and W. C. Steere, Eds.), pp. 157–193, Appleton-Century-Crofts, New York.

Spieth, H. T., 1974, Mating behavior and evolution of the Hawaiian *Drosophila*, in: *Genetic Mechanisms of Speciation in Insects* (M. J. D. White, Ed.), pp. 94–101, Australia and New Zealand Book Co., Sydney.

Spieth, H. T., 1978, Courtship patterns and evolution of the *Drosophila adiastola* and *planitibia* species subgroups, *Evolution* **32:**435–451.

Spieth, H. T., 1981, *Drosophila heteroneura* and *Drosophila silvestris*: Head shapes, behavior and evolution, *Evolution* **35:**921–930.

Spieth, H. T., 1982, Behavioral biology and evolution of the Hawaiian picture-winged species group of *Drosophila*, in: *Evolutionary Biology*, vol. 14 (M. K. Hecht, B. Wallace and G. T. Prance, Eds.), pp. 351–437, Plenum Press, New York and London.

Stearns, H. T., 1966, *Geology of the State of Hawaii*, Pacific Books, Palo Alto, CA.

Tarr, C. L., and Fleischer, R. C., 1993, Mitochondrial–DNA variation and evolutionary relationships in the amakihi complex, *The Auk* **110(4):**825–831.

Tarr, C. L., and Fleischer, R. C., 1995, Evolutionary relationships of the Hawaiian honeycreepers (Aves: Drepanidinae), in: *Hawaiian Biogeography: Evolution on a Hot Spot Archipelago* (W. L. Wagner and V. A. Funk, Eds.), pp. 147–159, Smithsonian Institution Press, Washington, D.C.

Templeton, A. R., 1979, Once again why 300 species of Hawaiian *Drosophila*? *Evolution* **33:**513–517.

Templeton, A. R., 1980, The theory of speciation via the founder principle, *Genetics* **94:**1011–1038.

Templeton, A. R., 1981, Mechanisms of speciation—A population genetic approach, *Annu. Rev. Ecol. Syst.* **12:**23–48.

Templeton, A. R., 1982, Genetic architectures of speciation, in: *Mechanisms of Speciation* (C. Barigozzi, Ed.), pp. 105–121, Alan R. Liss, Inc., New York.

Thomas, R. H., and Hunt, J. A., 1991, The molecular evolution of the alcohol dehydrogenase locus and the phylogeny of Hawaiian *Drosophila*, *Mol. Biol. Evol.* **8:**687–702.

Throckmorton, L. H., 1966, The relationships of the endemic Hawaiian Drosophilidae, *Univ. Texas Publ.* **6615:**335–396.

Val, F. C., 1977, Genetic analysis of the morphological differences between two interfertile species of Hawaiian *Drosophila*, *Evolution* **31:**611–629.

Wagner, W. L., 1991, Evolution of waif floras: A comparison of the Hawaiian and Marquesan archipelagos, in: *The Unity of Evolutionary Biology, Proc. Fourth Int. Congr. Syst. &*

Evol. Biology, Vol. 1 (E. C. Dudley, Ed.), pp. 267–284, Dioscorides Press, Portland, OR.

Wagner, W. L., Weller, S. G., and Sakai, A. K., 1995, Phylogeny and biogeography in *Schiedea* and *Alsinidendron* (Caryophyllaceae), in: *Hawaiian Biogeography: Evolution on a Hot Spot Archipelago* (W. L. Wagner and V. A. Funk, Eds.), pp. 221–258, Smithsonian Institution Press, Washington, D.C.

Walker, G. P. L., 1990, Geology and volcanology of the Hawaiian Islands, *Pacif. Sci.* **44:**315–347.

Watson, G. F., 1979, On premating isolation between two closely related species of Hawaiian *Drosophila*, *Evolution* **33:**771–774.

2

Codon Bias and the Context Dependency of Nucleotide Substitutions in the Evolution of Plastid DNA

BRIAN R. MORTON

Abbreviations: Ts—transition, tv—transversion, indel—insertion/deletion

Key Words: chloroplast, *rbcL*, *psbA*, molecular evolution, rate heterogeneity, context, neighboring base

INTRODUCTION

A great deal of the work in plant molecular evolution to date has focused on the chloroplast genome. This is due partially to certain advantages, such as very low levels of polymorphism and a lack of recombination, which simplify data manipulation, but also to the technical ease of working with chloroplast DNA (cpDNA). A bias towards studying cpDNA is most apparent in plant molecular systematics where it has long been the molecule of choice (e.g., Chase *et al.*, 1993; Clegg, 1993), but it has also been utilized in analyses of variation in evolutionary rates between land plant lineages (Gaut *et al.*, 1992), between loci (Muse and Gaut, 1997) and among

BRIAN R. MORTON • Department of Biological Sciences, Barnard College, Columbia University, New York, New York 10027.

Evolutionary Biology, Volume 31, edited by Max K. Hecht *et al.* Kluwer Academic / Plenum Publishers, New York, 2000.

different genomes (Gaut *et al.*, 1996; Eyre-Walker and Gaut, 1997). In addition, the chloroplast genome has been the subject of comparative genome organization studies (Downie and Palmer, 1992) and analyses of evolutionary processes at the molecular level (Curtis and Clegg, 1984; Ritland and Clegg, 1987; Delwiche and Palmer, 1996).

All of this work, particularly the research in systematics, has led to a vast accumulation of sequence data from the plastid genome of a diverse array of lineages. A total of 11 complete genome sequences are currently available, with more nearing completion (Palmer, 1993) and the ribulose bisphosphate carboxylase/oxygenase (*rbcL*) locus is one of the most widely sequenced genes; over 1,500 species are now represented in GenBank. Whereas interest in plant molecular evolution is expanding to encompass the nuclear and, to a lesser degree the mitochondrial genome, we are now in a position to learn a great deal about molecular evolution from this large data set, which has yet to be fully exploited.

This chapter reviews some of the recent studies concerning the molecular evolution of the plastid genome. Two different features are covered. The first is the evolution of codon usage bias in different lineages and what this might tell us about the evolution of the genome itself. The second is the complex pattern of nucleotide substitution that apparently exists in cpDNA. The data indicate a process with several levels of site-by-site interactions that could potentially influence how we understand and analyze evolution at the molecular level. In addition, both of these subjects may have importance for the controversies surrounding phylogenetic investigations of the plastid origins (Lockhart *et al.*, 1992) and for comparative genome analyses including systematics. Finally, observations pointing to areas of research that need to be addressed with future work are also discussed.

THE PLASTID GENOME

The complete plastid genome has now been sequenced for 11 different organisms spanning four phyla of the plant kingdom. These species are listed in Table I, along with *Chlamydomonas reinhardtii* that, although not yet fully sequenced, has been the focus of a great deal of work and that is discussed in the current study. One result of accumulating all of this sequence data is that a great deal is now known about the origins, structure, and coding content of this genome. In particular, the endosymbiotic theory, which proposes that the plastid arose from the endosymbiosis involving an ancestor of the present day cyanobacteria, has become well

TABLE I.   Species for which a Complete Plastid Genome Sequence is Available

Organism	Abbreviation	Genome AT Content	Noncoding AT Content	Accession
Cyanophora paradoxa	Cpa	69.5	80.9	U30821
Chlamydomonas reinhardtii[a]	Cre	—	—	—
Euglena gracilis	Egr	73.9	82.4	X70810
Chlorella vulgaris	Cvu	68.4	75.5	AB001684
Odontella sinensis	Osi	68.2	80.7	Z67753
Porphyra purpurea	Ppu	67.0	77.9	U38804
Marchantia polymorpha	Mpo	71.2	84.3	X04465, Y00686
Pinus thunbergii	Pth	61.5	66.2	D17510
Epifagus virginiana	Evi	64.0	74.6	M81884
Nicotiana tabacum	Nta	62.2	69.6	Z00044, S54304
Zea mays	Zma	61.5	69.4	X86563
Oryza sativa	Osa	61.0	67.8	X15901

[a] The complete genome sequence is not, yet available for *Chlamydomonas* but this species is discussed in the current study.

established (Reith, 1995; Palmer and Delwiche, 1996). The current data point to a single original endosymbiotic event but there is strong evidence that multiple secondary endosymbioses have occurred (Palmer, 1993; Reith, 1995; Palmer and Delwiche, 1996). This original endosymbiosis also apparently gave rise to the cyanelle of *Cyanophora paradoxa* (Palmer, 1993), so all general reference to plastids in this study include this organelle.

The plastid genome is a circular molecule that generally ranges in size from 100 to 200 kilobases (kb) in length. Variation in genome structure and coding content has been reviewed elsewhere (e.g. Palmer, 1991; Reith, 1995) but a short discussion will be helpful. The genome typically codes on the order of 120 to 250 genes, most of which have been identified. The majority of plastid genes have products involved in either photosynthesis or in gene expression (Reith, 1995) and, although there is a great deal of conservation of gene content among the different genomes, some variation exists. The most noticeable exceptions are parasitic plants, such as *Epifagus virginiana* that has a genome size of only 70 kb, approximately half the size of most land plants, and that has lost all genes involved in photosynthesis (Wolfe *et al.*, 1992). Although most plastid genes have been identified, there are a few open reading frames (ORFs) of unknown function. Several of these are conserved in two or more genomes and have been designated *ycf* (Hallick and Bairoch, 1994). These are treated as functional genes.

One important aspect for the purposes of the current study is the expression of plastid-encoded genes. All plastid genes are translated within the stroma by a prokaryoticlike ribosome composed of proteins coded by both the plastid or nuclear genome, and rRNA coded by the plastid. The plastid genome also contains either 30 or 31 tRNA genes, depending on the particular species (Palmer, 1991), which are utilized in translation of plastid genes. In terms of anticodon sequence, the plastid genomes sequenced to date have essentially the same tRNA content, or population, that is given in Table II by listing the complementary codons for *Cyanophora*.

Why the plastid genome retained the particular genes that it did following endosymbiosis has been the subject of some debate (Palmer, 1997). The current discussion is particularly relevant to why the plastid genome has the tRNA content seen in Table II. A certain pattern exists in the plastid tRNA genes. First, two- and threefold degenerate codon groups tend to be recognized by a single coded tRNA that is complementary to the A or C terminating codon depending on the particular group (UUC, CAC, GAC, AAC, UAC, UGC, AGC and AUC or AAA, CAA, GAA and AGA).

TABLE II. Codons with a Complementary tRNA Coded by the Cyanelle Genome of *Cyanophora*

		2nd Pos				
		U	C	A	G	3rd
	U	UUC	UCC	UAC	UGC	U
		UUA	UCA	Stop	Stop	C
		UUG	UCG	Stop	UGG	A
						G
1	C				CGU	U
s		CUC		CAC		C
t		CUA	CCA	CAA		A
					CGG	G
P	A					U
o		AUC	ACC	AAC	AGC	C
s			ACA	AAA	AGA	A
		AUG				G
	G					U
		GUC		GAC	GGC	C
		GUA	GCA	GAA	GGA	A
						G

Second, fourfold degenerate groups tend to have one or two tRNAs available and, again, it is usually the A or C terminating codons that are fully complementary. These two general rules appear to limit the number of tRNA genes while allowing either Watson–Crick bonding between the codon and anticodon or a G–U bond at the wobble position. The major exception to this rule is arginine (CGN) with tRNAs complementary to CGG and to CGT. This tRNA content is addressed in more detail in the following sections.

The current review focuses on two features of plastid DNA evolution. The first is the pattern of codon usage with reference to the tRNA genes, and the evolution of selection over time on codon usage in different lineages. The second is an examination of the pattern of nucleotide substitution, with respect to local base composition. Both of these areas of research suggest that the extensive data set available for plastid DNA will prove to be an excellent system to broaden our understanding of some basic processes of molecular evolution.

CODON BIAS

Codon Bias in Unicellular Organisms

A consideration of the nonuniform representation of synonymous codons, known as codon bias, has become a standard feature of DNA sequence analysis. Codon bias arises from an interplay between composition bias and selection, with different relative importances in different genomes and organisms (Sharp and Li, 1987b). Composition bias arising from a bias in the mutation process will tend to dominate in species with smaller effective population sizes, whereas selection will usually play a more significant role in those with larger population sizes, due to the relatively small selective differences between synonymous codons (Li, 1987).

The role of selection has been studied most thoroughly in *Escherichia coli* (Ikemura, 1985; Sharp, 1991). One of the fundamental observations concerning selection and codon bias is that there is a strong correlation between the relative representation of synonymous codons in highly expressed genes and the relative concentration of the cognate tRNAs in the cell (Ikemura, 1985). Overall, highly expressed genes have a strong bias toward a specific set of codons, called the major codons, that are complementary to abundant tRNAs, whereas low expression genes have a codon usage bias that is dominated by the genome composition bias (Andersson and Kurland, 1990; Sharp, 1991). This observation suggests a coadaptation of codon usage and tRNA abundancy (Andersson and Kurland, 1990) to reduce the effective number of codons, presumably increasing the efficiency of translation (Ikemura, 1985; Sharp, 1991).

Further evidence that is frequently cited in support of this model is the inverse correlation between synonymous substitution rate and the degree of codon bias that is observed in comparisons of homologous genes from *E. coli* and *Salmonella typhimurium* (Sharp and Li, 1987b; Sharp, 1991). This argument posits that highly expressed genes, those with the strongest codon bias, are under more stringent selection with the result that most silent changes are deleterious and constrained by negative selection. However, recent work examining the influence of selection for codon bias on silent substitution rate has drawn this simple relationship into question. First, Eyre-Walker and Bulmer (1995) observed that variation among *E. coli* genes in silent substitution rate exists for all synonymous codon groups, even those that show little or no variation in codon bias among genes. To explain the variation in rate among all codon groups they suggested that a transcription-coupled repair process could be responsible. In a second study, Berg and Martelius (1995) partitioned silent substitution rate variation among *E. coli* genes into variation in mutation rate and variation due to selection, although parts of their theory had been developed previously by Lewontin (1989). Consistent with the findings of Eyre-Walker and Bulmer (1995), they found that the variation in synonymous substitution rate was primarily a result of variation in mutation rate among genes.

Both of these studies indicate that, although there is a correlation between silent substitution rate and codon bias (Sharp and Li, 1987b), variation in mutation rate rather than selection on codon usage may be the causal factor. The theory shows that selection has to be quite strong to decrease the silent substitution rate, particularly when selection favors a codon that would not be highly represented as a result of composition bias alone (Eyre-Walker and Bulmer, 1995). Because selective differences between synonymous codons are small, selection favoring mutations to major codons in high-expression genes will be offset by a high rate of mutation away from major codons because most mutations will, by necessity, occur in a major codon due to codon bias (Berg and Martelius, 1995). Overall, both studies suggest that selection on codon usage may have little effect on synonymous substitution rate, a result that will have to be considered seriously in future work on codon bias.

A final feature of codon bias must be discussed. In studies of codon usage bias, one of the most frequently employed measures of the degree of bias, and the one that is used in this chapter, is the Codon Adaptation Index (CAI) of Sharp and Li (1987a). This measures, for a specific gene or codon usage table, the degree of adaptation toward a specified codon usage pattern that is considered "optimal." In studies of selection this optimal set is taken from the most highly expressed genes (Sharp and Li, 1987a), although any codon usage pattern could be used. Every codon within a synonymous group is assigned a fitness value from the optimal set based on its usage relative to the most highly represented codon in that group. The CAI value for any gene is a measure of the bias toward high-fitness codons. It should be stressed that CAI is not a strict measure of the degree of bias in the representation of synonymous codons but of the degree of adaptation to the optimal pattern that is chosen. A highly biased codon usage pattern can have a very low CAI value if it is strongly biased toward codons that are rare in the optimal set.

In the following sections the evidence that selection plays an important role in determining codon usage of plastid genes is discussed; low-expression plastid genes follow one pattern of codon usage whereas highly expressed genes are biased toward a different set of codons (the major codons) and the difference appears to be an adaptation of codon usage to the chloroplast tRNA content. Across all genes, a gradient is observed in the bias toward this second pattern and the gradient is correlated with expression level of the different genes, which strongly supports a model of selection to increase translation efficiency. It is also shown that selection intensity varies across lineages. Finally, the interesting relation between codon usage and rate of synonymous substitution in the plant *psbA* and *rbcL* genes is discussed.

Codon Usage Patterns in the Plastid Genome

The overall codon usage of different plastid genes follows the same basic rule as many unicellular organisms, such as *E. coli*. The basal codon usage pattern is generated by the genome composition bias and shows a bias toward the same nucleotides at the third codon position of every synonymous group, either A and T or G and C depending on the species. However, for each synonymous codon group there is a codon, the major codon, that is favored by selection, although in some codon groups the major codon matches the composition bias. Each gene has a codon usage that falls somewhere in the gradient from the basal codon usage to a high representation of major codons as a function of the expression level of that gene. This section discusses the composition bias and the major codons of the plastid genome before presenting evidence concerning codon bias and expression level.

All plastid genomes have a high AT content, although the flowering plants have lower AT biases than the algae (see Table I). This genome bias results in an increased representation of synonymous codons with an A or T at the third position. This is the basal pattern of codon usage for the plastid genome and it is observed in most plastid genes (Morton, 1993). Some genes, however, have a noticeably different pattern of codon usage. The second pattern is observed most frequently in certain genes of the various algal genomes; among the land plants, only the *psbA* gene from each genome shows this alternate codon usage pattern. The major codons of the plastid genome can be determined by a study of this alternate pattern of codon usage.

The unusual nature of the plant *psbA* gene relative to other plant chloroplast genes is apparent when various *Nicotiana* and *Cyanophora* genes are clustered by their similarity in codon usage (Fig. 1) based on distances measured using the method of Long and Gillespie (1991). The *Nicotiana psbA* gene clearly has a pattern that is unique for its genome but similar to what exists in a large number of *Cyanophora* genes. The cumulative codon usages of the *Nicotiana* and *Cyanophora* genes from Fig. 1 are given in Table III along with the codon usage of *psbA* from *Nicotiana*. The

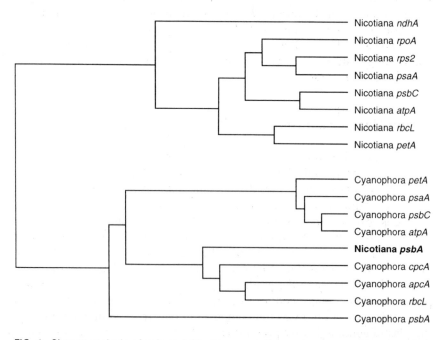

FIG. 1. Cluster analysis of selected *Nicotiana* and *Cyanophora* genes based on their similarity in terms of codon usage as measured by the method of Long and Gillespie (1991). Note the similarity between the *psbA* gene from *Nicotiana* (bold) and the *Cyanophora* genes.

TABLE III.　Codon Usage for *Cyanophora* and *Nicotiana* Plastid Genes from Fig. 1 and the *psbA* Gene from *Nicotiana*

	Codon	Cpa	Nta *psbA*	Nta		Codon	Cpa	Nta *psbA*	Nta
	CTG	1	1	24		GGG	2	1	49
	CTA	5	10	47	Gly	GGA	8	5	98
Leu	CTT	18	6	81		GGT	129	21	109
	CTC	0	0	18		GGC	6	6	35
	TTG	2	6	63					
	TTA	122	8	112		ATA	0	2	69
					Ile	ATT	80	17	132
	TCG	0	1	14		ATC	33	12	57
	TCA	3	2	26					
Ser	TCT	60	10	68	His	CAT	7	5	68
	TCC	13	4	41		CAC	19	6	25
	AGT	17	6	46					
	AGC	16	4	15	Gln	CAG	3	1	35
						CAA	65	5	89
	CGG	0	0	8					
	CGA	0	2	29	Glu	GAG	9	5	43
Arg	CGT	70	8	53		GAA	83	16	131
	CGC	5	2	15					
	AGG	0	1	17	Asp	GAT	56	3	109
	AGA	15	2	43		GAC	30	3	32
	CCG	3	0	19	Asn	AAT	17	9	84
Pro	CCA	33	5	35		AAC	45	13	34
	CCT	28	10	70					
	CCC	0	0	23	Lys	AAG	3	0	25
						AAA	55	0	118
	ACG	0	1	20					
Thr	ACA	13	2	49	Tyr	TAT	32	5	88
	ACT	74	10	69		TAC	30	7	19
	ACC	5	4	50					
					Cys	TGT	15	2	26
	GTG	0	0	24		TGC	2	0	6
Val	GTA	68	12	97					
	GTT	54	9	67	Phe	TTT	19	7	102
	GTC	0	0	18		TTC	50	19	59
	GCG	8	1	25					
Ala	GCA	89	7	87					
	GCT	93	24	118					
	GCC	0	3	43					

most noticeable difference between the codon usage of the two groups in Fig. 1 is the bias of twofold degenerate groups that have a pyrimidine at the third position (referred to from now on as the NNY groups) that are the groups coding for the amino acids His (CAY), Asn (AAY), Asp (GAY), Tyr (TAY), Cys (TGY), and Phe (TTY). The genome composition bias leads to a high representation of T at the third position (NNT codons) whereas in the *Cyanophora* genes represented in Fig. 1 and the *psbA* gene of *Nicotiana* there is an increased use of the NNC codons (see Table III) as well as the ATC codon in the threefold degenerate Ile group (Morton, 1993, 1996). The only exception to this increase is the lack of bias toward coding Cys with TGC. Therefore, in all NNY codon groups except TGY, the major codon is the NNC codon. For other codon groups, the similar bias in all genes indicates that the major codon does not differ from the composition bias.

Based on this difference between the two patterns it has been suggested that the second pattern is the result of selection adapting codon usage to the plastid tRNA content in order to increase translation efficiency (Morton, 1993, 1996). For each of the NNY groups, the NNC codon is fully complementary to the tRNA available for translation (see Table II) so the bias toward this codon, and away from the composition bias, results in a match between codon usage and the tRNA content. Because of this, the second pattern will be referred to as the selection pattern and will be used to define the major codons of the plastid.

Although both patterns have a high overall representation of T at the third position of fourfold degenerate codon groups, selection increases this bias further as demonstrated in Fig. 2. When the C content at the third position of the NNY codon groups is plotted against the T content of fourfold degenerate groups for *Cyanophora* genes, which tend to be under selection for codon usage as is discussed in the following paragraph, a significant correlation is seen. This correlation indicates that the major codons for the fourfold degenerate groups are the codons that have a T at the third position. This strong bias toward different third position nucleotides in different codon groups is difficult to reconcile with a model based strictly on mutation bias and provides good evidence for selection (Morton, 1996).

The major codons of plastid genes, those apparently favored by selection, raise an interesting question. Codon usage is apparently adapted to match the tRNA content in the NNY groups but not in the fourfold degenerate groups. In these groups, selection favors T at the third position whereas it is the A and C terminating codons that almost always have a complementary tRNA coded (Table II). Why would selection favor a match to the tRNA population in some cases but not others? A possible explanation is based on the suggestion that if selection on codon usage has its

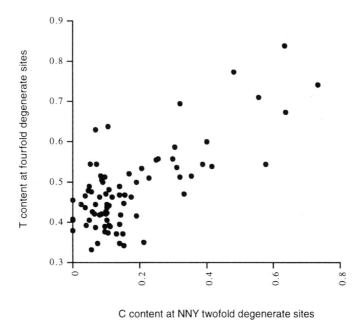

FIG. 2. A comparison of T content at the third position of fourfold degenerate codon groups to the C content at the third position of twofold degenerate NNY codon groups of *Cyanophora paradoxa* cyanelle genes.

basis in an increase in translation efficiency, then it may actually favor an intermediate codon/anticodon bond strength (Grosjean and Fiers, 1982). The bond should be neither too weak to prevent rapid incorporation nor too strong to result in a delay in tRNA release. The result would be that codons that are AT rich in the first two positions will be biased toward C over T at the third position whereas GC rich codons will be biased toward A or T when possible (Grosjean and Fiers, 1982; Grosjean, Houssier, and Cedergren, 1986). Although this model has not been supported in several studies (Andersson and Kurland, 1990), an appealing variation is the possibility that selection distinguishes between Watson–Crick bonding and wobble, as opposed to just GC content as suggested by Grosjean and Fiers (1982), as a function of the second position composition. In this case, codon usage would be adapted to match the tRNA content in codon groups with an A or a T at the second position, in order to increase bond strength, or away from full complementarity in groups with a G or a C at the second position to decrease the strength (Morton, 1996).

This model is supported to some extent by the exceptions to the general codon usage of the selection pattern. As already noted, the one

NNY group without NNC as the major codon is Cys (Morton, 1996; see Table III), which is coded by TGY as opposed to the others that have an A or a T at the second codon position. In addition, valine, coded by GTN, is an exception to the bias toward T at the third position of the fourfold degenerate groups (Morton, 1996; see Table III). However, it is also clear that this cannot account for every aspect of the selection pattern of codon usage. The high representation of AGC and CTT in *Cyanophora* (see Table III) does not fit the model and the extremely strong biases toward both GGT and CGT are very interesting. Overall, it remains unclear why selection favors the specific set of major codons that it apparently does in the plastid genome.

Although a model for how selection matches codon usage to the limited tRNA content of the plastid would be pleasing, it may be the case that codon bias in general is more complex. Specifically, it appears that the selection pattern of codon usage is not unique to the plastid. Instead, the highly expressed plastid genes appear to have a codon usage pattern that is very similar to what is observed in several other organisms. It has been suggested previously that there may be a set of codons that is favored in highly expressed genes from a number of organisms (Sharp *et al.*, 1988) and if this turns out to be true then a much more general model will be required.

Selection on codon bias has been studied in detail in *E. coli*, in *S. cereviseae*, and in *Drosophila* and in the plastid genome. In each case there is a strong bias toward the major codons in highly expressed genes, although evidence for selection remains somewhat controversial for *Drosophila* because expression levels are more difficult to establish. If we look at the relative representation of codons from specific codon groups in highly expressed genes, there is a striking similarity in codon usage between all four of these diverse genomes (Table IV). The AAC, ATC, and TAC codons are used preferentially as has been noted for *E. coli* and for yeast (Sharp *et al.*, 1988). However, there are other striking features, notably the virtual absence of ATA codons in all organisms and the lack of bias in the GAY group coding for Asp; for some reason, all of these organisms show nearly equal uses of GAT and GAC relative to the bias in the other NNY codon groups. Similarities are also observed in fourfold degenerate groups, in particular CGN and GGN where the lack of CGR and GGR codons in all organisms is quite obvious (Table IV). This occurs even though a purine at the third position is common in other fourfold degenerate groups (data not shown). Therefore, highly expressed plastid genes seem to be following a widespread pattern of codon usage.

The codon usages of the different organisms in Table IV are similar enough that it seems unlikely that they are converging by chance on this pattern of codon preference in highly expressed genes. A full under-

TABLE IV. Comparison of the Relative Codon Usages of Selected Codon Groups from Highly Expressed Genes of the Plastid Genome, *E. coli*, Yeast, and *Drosophila*

Codon	Plastid[a]	psbA	E. coli	Yeast	Drosophila
TAT	0.29	0.15	0.34	0.13	0.19
TAC	0.71	0.85	0.66	0.87	0.81
TTT	0.18	0.21	0.25	0.21	0.08
TTC	0.82	0.79	0.75	0.79	0.92
GAT	0.50	0.48	0.42	0.42	0.44
GAC	0.50	0.52	0.58	0.58	0.56
CAT	0.13	0.13	0.25	0.26	0.15
CAC	0.87	0.87	0.75	0.74	0.85
AAT	0.13	0.08	0.14	0.14	0.08
AAC	0.87	0.92	0.86	0.86	0.92
TGT	0.84	0.45	0.41	0.89	0.05
TGC	0.16	0.55	0.59	0.11	0.95
ATA	0	0	0.003	0.02	0
ATT	0.50	0.33	0.29	0.45	0.22
ATC	0.50	0.67	0.71	0.53	0.78
GGG	0.01	0	0.03	<0.01	0
GGA	0.05	0.05	0.01	<0.01	0.22
GGT	0.89	0.90	0.52	0.95	0.35
GGC	0.05	0.05	0.45	0.04	0.43
CGG	0	0	<0.01	0	0
CGA	0	0	<0.01	0	0.02
CGT	0.86	0.82	0.67	0.11	0.40
CGC	0.06	0.13	0.32	0	0.52
AGG	0	0	0	0.02	0.05
AGA	0.08	0.05	<0.01	0.87	0.01

[a] Relative codon usages from highly expressed plastid genes (this study), the *psbA* genes from the algae in Table I excluding *Euglena*, and high expression genes from *E. coli* (Lobry and Gautier, 1994), yeast (Sharp *et al.*, 1986), and *Drosophila* (Shields *et al.*, 1988).

standing of codon bias will have to be able to account for this convergence. Although almost every codon is a major codon in some organism (Andersson and Kurland, 1990), the question of whether there are certain sets of major codons that are optimal must be addressed in future work. If this is true, then the match between these codons and the plastid tRNA pool raises an interesting issue. Selection on codon bias probably involves a coadaptation between codon usage and tRNA concentration (Andersson and Kurland, 1990) that will lower the effective number of codons in highly expressed genes and match them to concentrated tRNAs. In the plastid, however, the codon usage is forced to adapt to the 31 tRNA genes; variation in concentration is not an option for many degenerate groups. If the codons favored in the highly expressed plastid genes are more universally favored by selection, then it suggests that those tRNA genes coded by the plastid genome were retained following endosymbiosis due to selection favoring a codon usage/tRNA pool that is optimal for translation efficiency. Therefore, the plastid genome may be evidence of a trade-off between translation efficiency and limiting the tRNA gene content. Given the similarities in codon usage in Table IV this possibility should be examined in future work.

Although the details of how codon usage is matched to tRNA content remain unclear for the plastid genome, a role for selection can be tested further. The codon usage of most genes from the completely sequenced plastid genomes reflects the high AT composition bias but every gene lies somewhere on the gradient from this basal codon bias to a high represen-tation of the major codons that were discussed previously. Every plastid genome has at least one gene with a noticeable selection pattern of codon usage with the exception of *Euglena*, which will be discussed separately in the following section. In an attempt to demonstrate more conclusively that selection acts on the codon bias, as has been suggested, we need to first measure levels of adaptation to the selection pattern, that is to the major codons, and then establish whether or not adaptation is correlated with known or expected expression level.

Selection on Codon Usage in Different Plastid Genomes

The first thing to be done is to estimate the relative strength of selec-tion in different plastid genes and in different genomes overall. All sequenced genomes will be tested with the exceptions of *Epifagus*, due to its drastically reduced coding content, and *Zea*, due to its close relationship to *Oryza*. The existence of a single set of major codons (i.e. the presumed selection pattern is the same for all plastid genomes), means that CAI can be measured against this pattern of codon usage (Morton, 1998). Because we are assuming for now that the selection pattern is an adaptation to increase translation efficiency, the *psbA* gene from each genome was selected as the basis for measurement for two reasons; first, this gene codes for the prominent translation product of the plant chloroplast (Mullet and Klein, 1987) and second, it is the one land plant chloroplast gene that has a noticeable selection pattern of codon usage (Morton, 1993). A CAI value was calculated for every plastid gene using as codon fitness values (see Sharp and Li, 1987a) the relative synonymous codon usages from the *psbA* gene of the same genome (see Morton, 1998). From this calculation, the codon usages of all genes with a CAI of 0.75 or more were combined and the relative codon frequencies of this cumulative codon table were used as fitnesses for a second round of CAI calculations. The result is that we can compare the relative adaptation to a single codon usage pattern that represents highly biased plastid genes (Morton, 1998).

The CAI values for a set of genes from each species, selected to give an accurate representation of the range that exists in each, are shown in Table V. The results for *Oryza* are essentially identical to those obtained for *Nicotiana* and are therefore omitted. Several things are apparent from this table. First, the algae tend to be much more strongly adapted to the selection pattern than the plants. Second, within the algae there is considerable variation within each genome, much more than the variation within the land plant genomes. Third, within the algae, *Cyanophora*, *Chlamydomonas*, and *Chlorella* appear to have stronger selection. *Chlamydomonas* in particular has much greater CAI values on average and much less variation than other algae suggesting that strong selection acts on many more genes. Finally, although there is no evidence of a selection pattern in any gene from *Euglena*, as mentioned previously, CAI values for some genes are at least comparable to the *psbA* gene of flowering plants that has an observable selection pattern (Morton, 1996). The variation in composition of the NNY codon groups among the same genes is given in Table VI. Within the land plants, *psbA* is noticeably unique, having a much larger C content than any other gene, including all *psbA* genes. Again, only *Euglena* fails to have a bias toward C in the NNY codons in at least one gene.

One potential difficulty in comparing CAI values directly is that the basal codon bias can differ among genomes due to variation in genome

TABLE V. CAI Values of Selected Plastid Genes from Different Organisms

Gene	Organism[a]								
	Cpa	Cre	Egr	Cvu	Osi	Ppu	Mpo	Pth	Nta
psbA	0.729	0.782	0.436	0.671	0.830	0.611	0.634	0.407	0.455
rbcL	0.704	0.754	0.484	0.645	0.739	0.478	0.514	0.334	0.363
psaA	0.518	—	0.365	0.414	0.447	0.368	0.417	0.286	0.279
psbC	0.585	0.711	0.409	0.472	0.500	0.349	0.427	0.270	0.281
atpA	0.625	0.725	0.343	0.439	0.520	0.383	0.501	0.267	0.296
petA	0.548	0.725	—	0.401	0.444	0.353	0.411	0.251	0.253
rps2	0.484	—	0.399	0.374	0.428	0.379	0.447	0.313	0.281
rpl14	0.350	0.578	0.320	0.315	0.353	0.266	—	0.236	0.253
rpoA	0.418	—	—	0.355	0.367	0.328	0.409	0.242	0.268

[a] See Table I for abbreviations used.

TABLE VI. C Content at the Third Codon Position of NNY Codon Groups from Selected Genes

Gene	Organism[a]								
	Cpa	Cre	Egr	Cvu	Osi	Ppu	Mpo	Pth	Nta
psbA	0.73	0.94	0.19	0.73	0.88	0.70	0.66	0.65	0.62
rbcL	0.64	0.88	0.18	0.52	0.60	0.33	0.25	0.31	0.34
psaA	0.32	—	0.13	0.29	0.30	0.28	0.14	0.33	0.27
psbC	0.40	0.76	0.13	0.35	0.38	0.29	0.14	0.32	0.34
atpA	0.35	0.59	0.21	0.23	0.30	0.21	0.16	0.20	0.22
petA	0.25	0.60	—	0.28	0.26	0.30	0.08	0.19	0.24
rps2	0.08	—	0.19	0.15	0.07	0.31	0.09	0.20	0.25
rpl14	0	0.11	0.23	0.26	0.09	0.13	—	0.26	0.26
rpoA	0.15	—	—	0.27	0.13	0.13	0.09	0.31	0.22

[a] See Table I for abbreviations used.

composition bias (see Table I). To overcome this potential difficulty we can test selection on individual genes directly by comparing the observed CAI to an expectation generated from the genome composition bias. For every gene, CAI is a function of nucleotide composition bias, selection, and, to a small degree, amino acid usage. This last possibility can be seen if we compare GAY and AAY from Table IV. A gene with all Asp residues (GAY) will be likely to use high-fitness codons solely because both codons have roughly equal (and high) values. On the other hand, a second gene with the same degree of codon bias but composed solely of Lys residues will have a lower CAI value because this gene will, by necessity, have a certain number of low-fitness codons. Therefore, we have to control for both nucleotide and amino acid composition in order to detect selection.

A separate test on every gene greater than 350 nucleotides in length was performed by generating a distribution of CAI from 100 random sequences (Morton, 1998). Each random gene was assigned the same amino acid composition as the actual gene (with sixfold degenerate amino acids treated as two separate groups, a fourfold and a twofold degenerate), and a codon usage was generated by assigning each codon a probability based on the nucleotide composition from the combined noncoding regions of that genome. The CAI value for each of the 100 random genes was then calculated and then used to generate a distribution of expected CAI for that gene. The actual CAI was then compared to the distribution by taking the first two moments.

The results for some representative genes from each genome are given in Table VII. A CAI value that is 2 or more standard deviations from the

TABLE VII. Comparison of Observed CAI to Randomly Generated
Distributions[a]

Gene	Species[b]								
	Cpa	Osi	Ppu	Cvu	Egr	Mpo	Pth	Nta	Osa
psbA	16.1	25.7	17.8	18.4	5.8	17.9	9.0	10.9	10.6
rbcL	22.1	21.5	9.4	14.9	7.4	9.9	4.5	4.9	4.5
psbC	14.4	9.7	4.9	11.6	4.4	5.8	2.3	***	***
rbcS	4.7	9.2	3.5	—	—	—	—	—	—
tufA	12.7	6.4	4.8	7.2	***	—	—	—	—
psaA	13.6	9.2	6.3	11.6	2.7	6.3	3.0	***	2.0
atpA	14.0	8.3	3.6	7.9	***	8.9	***	***	***
atpB	9.5	8.9	5.6	8.7	***	3.6	2.7	***	2.0
atpE	3.9	4.0	2.2	3.3	—	—	***	***	***
petA	7.2	3.7	***	4.5	—	2.7	***	***	***
petD	6.8	3.3	3.3	6.7	—	—	***	***	***
rpoA	2.4	***	***	2.4	—	***	***	***	***
rps2	4.3	***	2.0	2.1	***	2.4	***	***	***
rps3	5.0	***	***	4.3	***	***	***	***	***
rps4	3.7	***	***	2.9	***	***	***	***	***
rps8	2.8	***	***	***	***	—	***	***	***
rps11	2.5	***	***	***	***	***	***	***	2.1
rpl2	4.1	2.2	***	***	***	—	***	—	***
rpl14	***	***	***	***	***	—	***	***	***
rpl16	3.3	***	***	***	***	***	—	***	***

[a] Number of standard deviations above the mean of the random distribution. Values below 2
are designated by ***.
[b] See Table I for species abbreviations.

mean of the generated distribution is considered significant and is taken as
evidence that the observed codon usage cannot be explained by composition bias alone. As expected, a large number of genes in the algal genomes
show evidence for selection (Morton, 1998). In *Cyanophora*, 75 of 93 genes
gave a significant result, whereas 63% of *Chlorella* genes and 48% and 37%
of *Odontella* and *Porphyra* genes, respectively, gave a significant result. On
the other hand, very few genes from the flowering plants showed any evidence for selection; only the two highly expressed genes, *psbA* and *rbcL*,
show any strong significance. However, in *Marchantia* a large number of
genes (50% of the total) appear to be under selection. The results from
Pinus appear to be similar to *Nicotiana* and to *Oryza* in that selection is
apparent primarily in *psbA* and in *rbcL*. These results support those in Table
V and indicate that either selection has been drastically lessened within the
land plant lineage, particularly the vascular plants, or it has increased within
the algal lineages.

Codon Bias of *Euglena* Chloroplast Genes

Perhaps the most interesting aspect of Table VII is that significant results are obtained for *Euglena*, indicating that the relatively high CAI values of some *Euglena* genes (Table V) are, in fact, the result of selection. The genes *psbA*, *rbcL*, and *psbC* show significantly higher CAI than expected despite the fact, mentioned previously, that there is no evidence of a selection pattern for any gene, including *psbA*, in this genome (Morton, 1998). All *Euglena* chloroplast genes have a very strong bias toward A and T at the third codon position (Table VIII) and there is no indication that the NNC codons, which are major codons in other plastid geneomes, are increased in representation in any gene, including those with a significant result in Table VII (mainly *psbA*, *rbcL*, and *psbC*). What, then is the role of selection?

What is intriguing is that the three genes with a highly significant CAI value (Table VII) have the highest T content at the third position of four-fold degenerate groups (Table VIII). This suggests that the T content of fourfold degenerate groups is the main target of selection in *Euglena*, whereas twofold degenerate NNY codons are not affected. As expected, other algae have a strong positive correlation between NNY C content and CAI for individual genes (data not shown). However, there is a slight negative correlation in *Euglena* ($r = -0.19$) but a significant negative correlation between NNY C content and the T content at fourfold degenerate sites ($r = -0.560$), which is opposite to what is observed for *Cyanophora* (Fig. 2).

TABLE VIII.   Third Codon Position Compositions of Selected *Euglena gracilis* Plastid Genes

Gene	Composition	
	NNY groups C content[a]	Fourfold degenerate groups T content[b]
psbA	0.19	0.70
rbcL	0.18	0.67
psbC	0.13	0.64
psaA	0.13	0.60
atpA	0.21	0.38
rpoB	0.11	0.51
tufA	0.22	0.30
rps2	0.19	0.58
rpl2	0.11	0.56

[a] Proportion of twofold degenerate NNY codons with a C at the third position.
[b] Cumulative third codon position T content for fourfold degenerate codon groups.

Therefore, unlike in other species, the C content of NNY groups is either unaffected by selection or, more likely, is selected against, indicating that selection may have a different influence on codon usage in *Euglena* by favoring only a subset of the major codons. Why this should be the case, even though *Euglena* has the same tRNA content as other genomes (Hallick *et al.*, 1993), remains an interesting question to be addressed in research on plastid DNA codon bias.

The unique aspects of codon usage in *Euglena* are particularly interesting when its genome structure as a whole is considered. The chloroplast of *Euglena* is apparently the result of a secondary endosymbiosis and it is not absolutely required by the organism; individual *Euglena* cells from which chloroplasts have been removed can survive as heterotrophs (van den Hoek, 1995). In terms of genome structure it has a very low coding content, only 108 genes, but contains over 100 introns (Hallick *et al.*, 1993), compared with about 20 introns per land plant genome and only a single intron in *Cyanophora* (Palmer, 1991). How any of these features relate to the exceptional codon bias requires further investigation.

Codon Bias and Expression Level

Until now we have been assuming that the selection on codon usage indicated by the results in Table VII is to increase translation efficiency. This assumption has been based only on the unusual codon usage of the very highly expressed plant *psbA* gene relative to other plant chloroplast genes. In this section selection is tested more thoroughly by establishing that the variation in CAI is correlated with expression level as is observed in *E. coli*, in yeast (Sharp and Li, 1987b) and, apparently, in *Drosophila* (Shields *et al.*, 1988).

Among the land plants the *psbA* gene is consistently the only locus that shows evidence for a selection pattern (see Fig. 1) that has been used as evidence of selection for translation efficiency (Morton, 1993). The product of *psbA*, the D1 reaction-center protein of PSII, undergoes oxidative damage during light absorption, and as a result turns over at a very high rate, making it the most prominent translation product in chloroplasts (Mullet and Klein, 1987). Therefore, if there is selection based on expression level, the *psbA* gene is the one locus for which we would expect to find evidence.

To test other loci, plastid genes coded by at least two of the sequenced genomes were ranked by CAI in each genome separately and the average ranking per genome of every gene was then determined (Morton, 1998). The top 15 and bottom 15 genes of this ranking are given in Table IX. The

TABLE IX. Ends of the Distribution of Algal Plastid Genes When Ranked
by Codon Bias

Strong codon bias	Product[a]
psbA	PS II, D1 reaction-center protein
cpcB	Phycocyanin β subunit
rpl12	50S ribosomal protein L12
cpcA	Phycocyanin α subunit
apcB	Allophycocyanin β subunit
rbcL	Rubisco large subunit
rbcS	Rubisco small subunit
apcA	Allophycocyanin α subunit
tufA	Elongation factor Tu
petD	Cytochrome b6/f subunit IV
atpA	ATP synthase CF_1 subunit α
atpB	ATP synthase CF_1 subunit B
apcF	Allophycocyanin B18 subunit
psbD	PSII, D2 reaction-center protein
psbB	PSII CP47 chlorophyll protein
Weak codon bias	
rpoA	RNA polymerase α chain
psaD	PSI ferredoxin-binding protein II
dnaB	DNA-replication helicase
trpG	Anthranilate synthase component II
chlB	Protochlorophyllide reductase chlB chain
rpl16	50S ribosomal protein L16
secY	Preprotein-translocase subunit Y
psaL	PSI subunit XI
ycf5	Unknown
rpl4	50S ribosomal protein L4
rpl14	50S ribosomal protein L14
rpl2	50S ribosomal protein L2
ycf38	Unknown
rps13	30S ribosomal protein S13
ycf4	Unknown

[a] Based on Kowallik *et al.*, 1995.

expected correlation with expression level is clear in this ranking. All of the high-ranking genes code for proteins that exist in high concentration in plant chloroplasts, based on concentration in SDS gel electrophoresis (Klein and Mullet, 1986; Mullet and Klein, 1987). These include the core proteins of PSII, the subunits of rubisco (*rbcS* and *rbcL*), which is the most abundant protein in plants, the translation elongation factor Tu (*tufA* gene), and the major subunits of the ATP synthetase. In addition, the genes coding for the major phycobiliproteins (designated *apc* and *cpc*), components of the Light Harvesting System in some algae, are also ranked quite high. At the low end of the distribution are genes that code for proteins that are not present in high concentrations in chloroplasts as well as some conserved ORFs of unknown function. Therefore, there is a good correlation between expression level and average CAI ranking (Morton, 1998) and, although expression level is not necessarily the same as translation level it is likely to be a good indicator for many genes.

It is interesting that, unlike *E. coli* where ribosomal proteins are highly expressed and have relatively high CAI values (Lobry and Gautier, 1994), the *rpl* and *rps* genes tend to have relatively low CAI values in plastids. It is clear from gel electrophoresis that the photosynthetic proteins are present in much larger quantities (Klein and Mullet, 1986) but it is not obvious why plastids do not have high expression levels of ribosomal protein genes in addition to the photosynthesis genes. One possibility is that the relatively small coding content of the plastid genome results in a low overall level of translation in the organelle such that ribosome protein expression can be relatively low.

The overall correlation between codon adaptation and relative abundance of the gene products fits very well with the general model of selection on codon bias. The bias toward the selection pattern varies among genes as a function of expression level probably as the result of varying selection for translation efficiency. In this way the plastid genome appears to match what has been observed previously in a variety of unicellular organisms (Ikemura, 1985).

Maintenance of Codon Bias

Although all of the genomes presented in Table I are widely diverged at the sequence level, past saturation at synonymous sites (data not shown), composition features are strongly conserved within the algae. Various comparisons with *Cyanophora*, chosen because of the evidence for strong selection in this organism (Tables V and VII) and the complete genome sequence available, show that both CAI and NNY C content are strongly conserved in other algae (Fig. 3) with a significant correlation in most cases (data not shown). The only exception is the comparison to the NNY C content in *Euglena*, although it is clear that relative codon adaptation is strongly correlated. This again suggests that codon adaptation exists in *Euglena* but that it is not determined by adapting the NNY codons, as discussed previously.

When *Cyanophora* is compared to *Marchantia* significant correlations are observed, but when it is compared to *Pinus*, to *Nicotiana*, and to *Oryza*

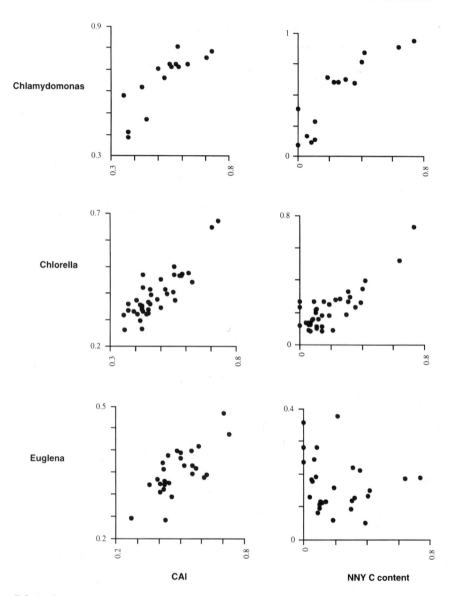

FIG. 3. Comparisons of *Cyanophora* genes to the homologous genes from each of the organisms listed in Table I except *Epifagus* and *Zea*. Both Codon Adaptation Index (CAI) and C content of NNY twofold degenerate codon groups are compared for genes over 350 nucleotides in length.

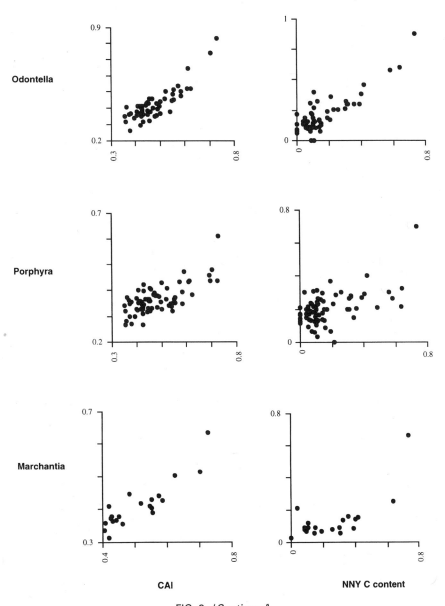

CAI

NNY C content

FIG. 3. (*Continued*)

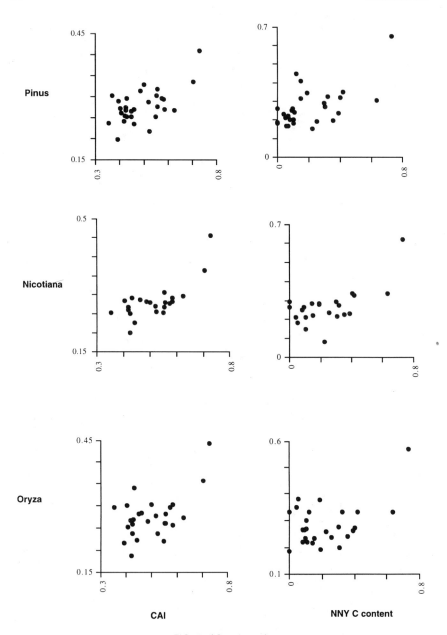

Pinus

Nicotiana

Oryza

CAI

NNY C content

FIG. 3. (*Continued*)

the correlations are weak and not significant if *psbA* is excluded (Morton, 1998). Therefore, the correlations exist in comparisons to lineages for which the random sequence generation gave a number of significant results (Table VII).

Codon Usage of the Plant *psbA* Gene

In the land plants, the pattern of codon usage in the plastid genome that is associated with selection is noticeable only in the *psbA* gene (see Figure 1) although there is good evidence (Table VII) for selection on other genes, in particular *rbcL*. Evidence for selection on *psbA* and *rbcL* is not surprising given the very high expression levels of these two genes, but in this section we examine these genes from the flowering plants in further detail. First, patterns in the synonymous substitution rate are discussed followed by the evolutionary dynamics of codon bias over time.

When the rates of synonymous substitution for *psbA* and for *rbcL* are compared to other flowering plant chloroplast genes greater than 750 nucleotides in length an interesting pattern emerges (Morton, 1997a). In a comparison of *Oryza* and *Nicotiana*, synonymous rates were estimated for different degeneracy groups separately using a variation of the method described by Lewontin (1989). Although *psbA* has the lowest overall rate of synonymous substitution (Morton, 1994), when genes are compared by rate at the NNY twofold degenerate groups, *psbA* has the highest rate of change and *rbcL* has the second highest. On the other hand, *psbA* has the lowest and *rbcL* the second lowest rate at fourfold degenerate sites. In the twofold degenerate NNR groups, *psbA* is about average in silent rate whereas *rbcL* is just slightly higher than average (Table X). Therefore, the

TABLE X.   Rates of Synonymous Substitution for Specific Degeneracy Groups from a Comparison of Rice and Tobacco

Gene	NNY rate[a]	NNR rate	Ile rate[b]	Fourfold rate
psbA	0.40 (0.17)	0.28 (0.03)	0.40	0.26 (0.09)
rbcL	0.34 (0.15)	0.34 (0.05)	0.53	0.47 (0.17)
Others[c]	0.24	0.31	0.22	0.56

[a] Synonymous rate calculated separately for NNY and NNR twofold degenerate codon groups and fourfold degenerate codon groups.
[b] Proportion of Ile codons with a synonymous difference.
[c] Average rates for 10 other chloroplast genes of length >750 nucleotides (Morton, 1997b).

two genes with significantly different codon usages also share an unusual pattern of substitution rate.

In addition to the unusual substitution rates, it has also been observed that the atypical pattern of codon usage observed in the flowering plant *psbA* gene appears to be degrading (Morton and Levin, 1997), and the same is true for *rbcL*. Analyses of the pattern of evolution over time at the molecular level have become more common recently, offering the advantage that it allows an examination of trends as well as an identification of selective sites (Yang, Kumar, and Nei, 1995; Stewart, 1995). A study of the evolution of a ribonuclease gene family found increases in both activity and thermostability when extant enzymes were compared to synthesized putative ancestral enzymes (Jermann *et al.*, 1995). In addition, the evolutionary dynamics of both lysozyme (Malcolm *et al.*, 1990) and chymase (Chandrasekharan *et al.*, 1996) have been studied by synthesis of enzymes using putative ancestral sequences.

To study the *psbA* locus the gene was sequenced for several angiosperm taxa that were then aligned with the outgroup taxa *Marchantia* and *Pinus*. A neighbor-joining tree (Saitou and Nei, 1987) using the Kimura 2-parameter model (Kimura, 1980) was generated using PHYLIP (Felsenstein, 1993) and the topology was used for a parsimony reconstruction of the ancestral angiosperm *psbA* sequence following the algorithm of Fitch (1971). For the putative ancestral sequence as well as the extant *psbA* sequences, the CAI of Sharp and Li (1987a) was calculated relative to three different codon usage patterns, all presumed to be selection patterns—the ancestral sequence itself, the *Marchantia psbA* gene, and the *Chlamydomonas psbA* gene (see Table V). Therefore, changes in CAI measure the change over time in the level of adaptation to the selection pattern of codon usage. In all flowering plant lineages, CAI of the *psbA* gene is lower than the adaptation of the inferred ancestral sequence (Morton and Levin, 1997; see Table XI). This indicates that adaptation to the selection pattern appears to have been degrading during angiosperm radiation. In addition, the results are robust regarding the topology used to reconstruct the ancestral sequence (Morton and Levin, 1997).

One of the major questions regarding any analysis utilizing ancestral reconstruction is the reliability of the reconstruction itself. The main difficulties for reconstruction, like any comparative molecular analysis, are multiple hits per site and parallel changes in different lineages. Both of these increase in probability with increasing divergence. When few changes have occurred reconstruction is a relatively simple matter, only when divergence increases is it necessary to apply a model of substitution to estimate the ancestral state (Yang, Kumar, and Nei, 1995). Therefore, it is not surprising that simulation analyses have found that the ancestral state for

TABLE XI. Evolution of CAI at the *psbA* Locus during Flowering Plant Divergence

Taxon	CAI-Anc[a]	CAI-Cre	CAI-Mpo
Ancestor	**0.733**	**0.378**	**0.617**
Hordeum	0.638	0.302	0.552
Secale	0.668	0.348	0.577
Oryza	0.637	0.318	0.538
Acorus	0.644	0.319	0.527
Phoenix	0.670	0.333	0.565
Ludisia	0.610	0.278	0.483
Arabidopsis	0.653	0.319	0.546
Brassica	0.675	0.346	0.573
Gossypium	0.638	0.325	0.551
Nicotiana	0.619	0.337	0.525
Petunia	0.623	0.336	0.524
Solanum	0.610	0.335	0.522
Spinacia	0.650	0.319	0.538
Vicia	0.603	0.291	0.512

[a] CAI was calculated relative to the codon usage of the ancestral sequence, *psbA* of *Chlamydomonas*, and *psbA* of *Marchantia*.

sequences that are less than 10% diverged is highly reliable (Zhang and Nei, 1997) regardless of the methodology (Hillis, Huelsenbeck, and Cunningham, 1994).

In the case of *psbA* we have sequences that fit this condition for accurate reconstruction. The largest branch length estimated by PHYLIP was 0.042 substitutions per site and all but one are less than 0.025 (Morton and Levin, 1997). Therefore, sequences are well under the divergence level found by Zhang and Nei (1997) and we can expect high reliability.

The same analysis performed on 60 flowering plant *rbcL* sequences spanning all major angiosperm groups also showed a lower CAI value in all 60 extant genes relative to the putative ancestral sequence (data not shown). As with *psbA*, the *rbcL* sequences used are not highly diverged and the reconstruction is expected to be reliable. The analyses of putative ancestral sequences indicate strongly that the two genes that show the unusual rate variation (Table X) have been decreasing in their adaptation to the selection pattern of codon usage during the divergence of the flowering plants.

How do we explain both the pattern of rate variation and the evolution of CAI for *rbcL* and for *psbA*? An appealing hypothesis is that selection has been an important factor determining codon usage in the past but that it was recently relaxed (Morton, 1997a; Morton and Levin, 1997), perhaps just prior to the divergence of the flowering plants. Both the rate patterns and the evolution of CAI could be explained by this model. The evidence for selection from Table VII could simply represent the remnants of a bias that has not been degraded by the mutation bias. The rate results are not as simple. Although it has been generally held that selection on codon usage will decrease synonymous substitution rate (e.g., Sharp and Li, 1987b), recent theoretical work shows that selection has to be fairly strong to get a reduction in rate (Eyre-Walker and Bulmer, 1995). For weak selection, no decrease in selection is expected and under certain conditions it can actually increase synonymous rate. This will occur when there is a strong mutation bias and counterselection (Eyre-Walker and Bulmer, 1995). Although this could potentially explain the rate results in Table X, the mutation bias measured for the chloroplast (Morton and Clegg, 1995) are not strong enough to generate the increase in rate that is observed (data not shown). It is also possible, of course, that selection was relaxed but is not completely absent, only reduced. What remains to be determined is whether or not the rate pattern from Table X can be explained by mutation bias alone as suggested by the codon usage pattern of *psbA* and the high rate of replacement of C and G in cpDNA (Morton, 1997a).

Testing this model further will require new theory. The model of Berg and Martelius (1995) measures selective pressure on twofold degenerate codon groups as opposed to mutation pressure and is one potential approach. The disadvantage, however, is that it uses a measure of codon bias taken from the gene under question to estimate selection, which is done by comparisons to low-expression genes. This is an implicit assumption that the gene is stationary, meaning that it is inadequate for tests of this very point, as with the model for *psbA*. In addition, models based on comparing polymorphisms to fixations (Akashi, 1995) are not applicable because polymorphic sites are virtually absent from cpDNA. Therefore, we need to develop a theory to test genes that are in the process of evolving from one stationary state to another.

Evolution of Selection on Codon Usage

If we are to consider a model in which selection on the codon usage of *psbA* has been recently relaxed, as suggested by the decreasing CAI values, then we have to consider reasons for why such a relaxation would have occurred. The most promising model so far is that there was a gradual decrease in selective pressure on codon usage during the evolution of the land plants. Prior to their divergence, it is likely that a situation similar to what is observed in *Cyanophora* and in *Chlamydomonas* existed, with the selection pattern apparent in a number of highly expressed genes. As land plants diverged, the selection pattern of codon usage would have gradually disappeared as selection was relaxed.

The promising aspect of this model is that if selection intensity gradually decreased over time then selection on codon usage would be relaxed on different genes at different times. As a result, selection would have persisted longest on highly expressed genes and may only have been relaxed quite recently. This would explain the existence of the strongest selection pattern in *psbA*, as well as the apparently weak adaptation toward the pattern in the *rbcL* gene (Table VII), if selection was relaxed most recently in both of these genes, meaning that they have not yet reached a stationary state. To test this further, it will be necessary to generate a large enough data set for a gene that is stationary but also not highly diverged among the angiosperm to allow for accurate reconstruction.

If this model is to be acceptable then we need to understand why selection would gradually decrease over time. Two potential factors could generate this result. One is a gradual increase in genome copy number over time (Morton, 1997a). Because *Chlamydomonas* has a single chloroplast per cell (van den Hoek, 1995) whereas flowering plants have multiple organelles, it is likely that an increase in genome copy number per cell has occurred during land plant divergence. The result of such an increase would be that selection for translation efficiency/rate could be replaced by the availability of an increased number of transcripts. The second factor is effective population size. If population size has decreased during the emergence of flowering plants then the effectiveness of selection on codon usage could disappear. The low selective values distinguishing synonymous codons, on the order of 10^{-9} (Hartl, Moriyama, and Sawyer, 1994) suggests that it is only effective in species with very large population sizes (Li, 1987).

The importance of distinguishing this model from maintenance by selection is that it has broader implications for the plastid genome. If the model is correct then it raises the possibility that the entire flowering plant chloroplast genome is in a state of change from one composition pattern to another. Of most interest would be the relation to overall AT content. In general, flowering plants have lower AT content than algae and *Marchantia* (see Table I) and this difference seems to be correlated with a much lower level of selection on codon usage (see Table VII). It will be of interest to investigate whether or not the two factors are related because the relaxation of selection on codon usage might be coincident with a relaxation of selection on AT content, suggesting that the AT content, of the plant genome is currently decreasing.

The results also raise questions about the suitability of cpDNA for certain types of analysis. If the genome, or even just specific genes, are not stationary then it would be important to consider this in phylogenetic analyses and in studies that date plant evolution by considering rates of change. Certainly, the *psbA* gene is a very poor choice for such analyses and even the *rbcL* gene may not be as appropriate as other plastid genes. Finally, the potential influence of selective differences between synonymous codons should be extended to analyses of selective differences between nonsynonymous codons as a function of translation efficiency that could have significant influences on the amino acid usages of proteins. A full understanding of protein structure will require a knowledge of these factors.

MUTATION DYNAMICS

Models for Nucleotide Substitution

One of the basic requirements for the molecular evolutionary analysis of any gene or genome is an understanding of the underlying nucleotide substitution process. An appropriate substitution model is a central feature of most molecular comparative studies, including methods of estimating sequence divergence, phylogenetic reconstruction that frequently assigns unequal weights to different events based on observed frequency (Swofford and Olsen, 1990), and sequence alignments (Thorne and Churchill, 1995). The widespread application of cpDNA in plant molecular evolution as well as the issues raised by the evolution of codon bias at the *psbA* locus require that we develop a solid understanding the basic substitution dynamics of cpDNA. In the following sections, the substitution process of cpDNA is discussed with a focus on the complex site interdependencies that have become apparent from recent analyses.

Generally, the evolution of DNA is modeled as a Markov process (Lewontin, 1989) with the simplifying assumptions that the transition matrices for different sites are identically distributed and that positions are independent (Kelly, 1994). As a result, differences in substitution dynamics among sites are assumed to result from selection. Several different models of substitution have been proposed based on observations of noncoding DNA or pseudogene sequences. It has become well established that different substitution types occur with different frequencies, with transitions usually observed to occur at a much higher frequency than transversions (Brown *et al.*, 1982; Gojobori, Li, and Graur, 1982; Li, Wu, and Luo, 1984). The simplest, and possibly the most commonly utilized, model that accounts for this bias is the Kimura 2-parameter model in which transitions occur at one rate and all transversions at another (Kimura, 1980). In addition to this, many models have been developed to account for various other substitution patterns, particularly different rates of the various transversions, extending to those that account for a different rate for every substitution (Nei, 1987). However, all of these models make the two simplifying assumptions stated previously.

Although the plastid genome has been widely utilized in plant molecular evolutionary studies, and has come to dominate molecular plant systematics, relatively few analyses of the substitution process in cpDNA have been performed. Early work noted that it is a conservative genome (Curtis and Clegg, 1984) that evolves on average at a rate severalfold lower than nuclear genes although not as slowly as plant mitochondrial DNA (Wolfe, Sharp, and Li, 1987). The most thorough analysis of cpDNA was performed by Ritland and Clegg (1987), who tested models of evolution among different chloroplast lineages and genes. They found that for third codon positions, synonymous substitutions were biased toward transitions and both reversibility as well as equal frequencies for complementary bases were rejected. The general bias toward transitions was also noted by Zurawski and Clegg (1987), who concluded that it had persisted over a long period of time in higher plant evolution.

Neighboring Base Composition and Substitution Bias

Despite the fact that simplifying assumptions are made in all of the models discussed previously, few studies have attempted to test them rigorously. However, there is some evidence that intersite dependencies, or influences of neighboring bases, can significantly affect the substitution process. Certainly, in the case of RNA coding sequences, selection on secondary structure can lead to significant correlation among sites that can influence studies of sequence evolution (Muse, 1995).

In addition to selection, an influence of context on the properties of mutation in DNA (Topal and Fresco, 1976) can generate nonindependence among sites. Many *in vitro* studies have provided evidence that replication and repair are both context dependent processes (Petruska and Goodman, 1985; Radman and Wagner, 1986; Jones, Wagner, and Radman, 1987; Mendelman *et al.*, 1989) and comparative analyses of substitutions in mammalian pseudogenes have revealed strong differences in substitution rate for different types in different contexts (Blake, Hess, and Nicholson-Tuell, 1992; Hess, Blake, and Blake, 1994).

Recent analyses of nucleotide substitutions in cpDNA have uncovered evidence for a process that is affected by complex set of interactions between sites that generates significant variation in neutral substitution dynamics. A basic level of variation is that the proportion of substitutions that are transversions varies widely among noncoding regions in the flowering plant chloroplast genome as a function of the AT content of the regions (Morton, 1995). Comparisons of rice and maize over a number of noncoding regions (described in Morton, 1995), aligned using the maximum likelihood method of Thorne and Churchill (1995), show a wide variation in Tv/Ts (see Fig. 4) and this variation is significantly correlated with the AT content of the different regions.

It should be noted for potential analyses of other genomes that the use of the ML alignment algorithm has proven critical as it allows for the exclusion of aligned regions that fall below a statistical threshold and, therefore, are not likely to be homologous, a procedure that standard alignment algorithms do not perform. Because noncoding regions often undergo a high rate of indels, sometimes overlapping in different lineages, alignment of nonhomologous stretches is a serious concern. A similar analysis of the same noncoding regions using alignments generated by the program CLUSTALV shows no correlation between Tv/Ts and AT

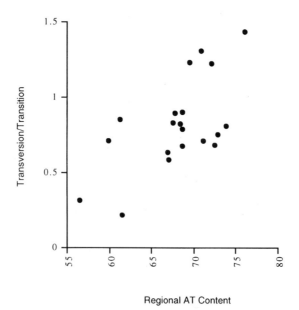

FIG. 4. A comparison of the AT content (of the entire region) to substitution bias at fourfold degenerate sites, for various coding and noncoding segments of the chloroplast genome from grasses.

content due to the relatively large number of random differences in non-homologous aligned stretches that are interpreted as substitutions (data not shown).

One possible explanation for the correlation in Fig. 4 is that A and T undergo transversions at a higher rate. This is not upheld by the observation that coding sequences fall onto the same regression if Tv/Ts of four-fold degenerate sites, which is very low relative to noncoding sequences, is plotted against AT content of the entire gene (Morton, 1995). When Tv/Ts is plotted against the AT content of only the fourfold degenerate sites, coding sequences fall onto a very different regression (data not shown). Therefore, the variation in Tv/Ts among all regions, coding and noncoding, is a function of the AT content of the entire region, not the composition of the bases actually undergoing substitutions.

The correlation between regional AT content and substitution bias suggests that local composition has a significant influence on substitution dynamics (Morton, 1995). This interpretation is supported by comparisons of substitution type to the average base composition of nucleotides near the site of substitution. Because of the low divergence of rice and maize at the nucleotide level (less than 10% difference in aligned regions; Morton, 1995) multiple hits are not a confounding problem and it is also a simple matter to define the composition of surrounding nucleotides for each substitution. In Table XII, substitution bias is compared to the AT content of

TABLE XII.   The Relationship between Transversion Proportion and the Composition of Flanking Nucleotides

	AT content[b]			
Distance[a]	0	1	2	χ^2
1	0.242	0.377	0.524	$p < 0.001$
2	0.327	0.384	0.493	$p < 0.025$
3	0.302	0.385	0.467	$p < 0.05$
4	0.283	0.383	0.457	$p < 0.05$
5	0.386	0.452	0.412	NS
6	0.422	0.419	0.432	NS
7	0.479	0.431	0.429	NS
8	0.429	0.424	0.410	NS
9	0.486	0.392	0.461	NS
10	0.442	0.420	0.450	NS

[a] Distance from the site of substitution of the two nucleotides, one 5′ and one 3′, that are used to measure AT content.
[b] AT content of the two bases on either side of the site. For each comparison the proportion of substitutions that are transversions is given.

specific nucleotide pairs that are increasingly removed from the substitution site. Each pair represents one base that is 5' and one that is 3', both the same number of nucleotides away from the site of substitution.

In the comparison of rice and maize the strongest correlation between substitution bias and neighboring base composition exists for the two bases that immediately flank the substitution site as reported previously (Morton and Clegg, 1995; Morton, 1995). This has also been recently reported for a noncoding region sequenced from the dicot group Celastrales (Savolainen, Spichiger, and Manen, 1997). The correlation has also been observed in a comparison of coding sequences, both in a pairwise comparison of 19 coding regions from rice and maize and a large-scale analysis of *rbcL* from a number of flowering plant lineages (Morton, 1997b).

Although the composition of the two flanking bases has the strongest correlation, there is also a significant correlation between substitution bias and composition of nucleotide pairs up to four sites removed from the substitution. In all cases, increasing AT content is related to increasing transversion proportion. This correlation between the composition of distant sites and substitution bias supports the observation that the overall AT content of several nucleotides surrounding a substitution is significantly correlated with substitution bias (Morton, 1997b; Morton, Oberholzer, and Clegg, 1997).

The influence of flanking nucleotides other than the immediate neighbors on substitution bias, however, is not straightforward. Rather, it appears to be dependent on the composition of the two immediate neighbors (Morton, 1997b; Morton, Oberholzer, and Clegg, 1997). In Fig. 5, substitution bias is compared to the AT content of 10 neighboring bases. This is done separately for sites with each of the three possible AT contents of the two neighboring bases. When the two immediate neighbors are both A and/or T there is a significant increase in the transversion proportion as AT content over 10 bases increases ($\chi^2 = 25.5$, $p < 0.01$), whereas in the other two cases (either one or two flanking G/C) the variation in substitution bias is not significant. The dominant importance of the composition of the two immediate neighbors was also demonstrated by Morton, Oberholzer, and Clegg (1997).

In addition to the influence of local AT composition on substitution bias there is a secondary effect of pyrimidine content. Within the noncoding regions in rice and maize, a 5' pyrimidine is found to result in a significant increase in the rate of transversion substitutions (Morton, Oberholzer, and Clegg, 1997). This result is cumulative so that YXR sites, where X represents the site of substitution, having two 5' pyrimidines (one on each strand) have the highest proportion of transversion and RXY sites have the lowest. This influence of an upstream pyrimidine is secondary to AT content

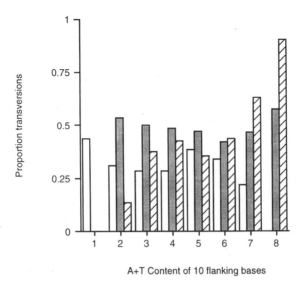

FIG. 5. Average substitution bias at a site and the AT content of the 10 neighboring bases, five 5′ and five 3′. The three plots represent sites AT contents of the two immediately flanking bases of 0 (open), 1 (shaded), and 2 (hatched). The variation in substitution bias is significant only in the latter case (see text).

such that the bases are ranked T > A > C > G with respect to the proportion of transversions observed. Overall, as a result, the context TXA has a strong bias toward transversions (62.4% of substitutions) in noncoding regions whereas the context GNC has a much lower proportion (23.1%).

This effect of pyrimidines is also observed in a comparison of 19 coding regions from rice and maize but it is more difficult to compare contexts at fourfold degenerate sites due to restrictions created by the genetic code. However, as in the noncoding DNA, the CXG context—where X is a fourfold degenerate third codon position—has the highest transversion proportion when both flanking bases are G and/or C and in those cases where the second codon position is G or C and the 3′ neighbor is an A or a T, there is a significantly higher proportion of transversions when that 3′ neighbor is an A (data not shown). Further, in the codon groups CTN and GTN (Ser and Val), 63.2% of substitutions upstream of an A are transversions whereas only 25.0% of substitutions upstream of a T are transversions. Therefore, in every case, a 5′ pyrimidine is associated with an increased proportion of transversions when AT content of the two flanking bases is the same.

The results demonstrate that there is an overlapping set of factors that influence substitution bias at apparently neutral sites in cpDNA of flowering plants. The most influential factor appears to be the AT content of the two flanking bases. As a function of this content, the composition of a slightly larger region is influential. In addition to the influence of AT content, an upstream pyrimidine increases transversion proportion.

Context and Synonymous Substitution Rate

Estimating substitution rate variation has been an important issue in molecular evolution and molecular systematics, since it has been demonstrated that rate heterogeneity has a significant influence on the accuracy of tree reconstruction (Jin and Nei, 1990; Huelsenbeck and Hillis, 1993; Tateno, Takezaki, and Nei, 1994; Mindell and Thacker, 1996; Yang, 1996) and other estimates, such as substitution bias (Wakeley, 1994). Methods have been developed to account for rate variation among sites in phylogenetic reconstruction using maximum likelihood (Yang, 1994; Felsenstein and Churchill, 1996), neighbor-joining (Jin and Nei, 1990; Tateno, Takezaki, and Nei, 1994) or differential weighting in parismony (Mindell and Thacker, 1996). In addition, several methods have been developed to estimate a parameter of variation when the substitution rate is assumed to follow a gamma distribution among sites (Gu, Fu, and Li, 1995; Yang and Kumar, 1996; Gu and Zhang, 1997). None of these methods make prior assumptions about the rate of evolution at specific sites, although rates can be reduced in number and assigned to specific codon positions (Felsenstein and Churchill, 1996).

When we consider cpDNA, there has been some effort to develop a better understanding of the evolutionary processes at the *rbcL* locus (e.g., Albert *et al.*, 1994), including rate heterogeneity (Kellogg and Juliano, 1997), due to the wide utilization of this gene in systematics. Studies of rate variation, though, have focused on selection as the underlying cause, particularly constraints on amino acid changes, and no real consideration has been given to understanding synonymous rate variation among sites. However, it is possible that variation in context, in terms of base composition of neighboring nucleotides, generates significant variation in synonymous rate. In mammalian pseudogenes, substitution rate is influenced by neighboring base composition (Blake, Hess, and Nicholson-Tuell, 1992; Hess, Blake, and Blake, 1994) and if context has a significant influence on synonymous substitution rate in cpDNA then this additional source of heterogeneity could prove to be as important as amino acid replacement heterogeneity (Kellogg and Juliano, 1997). In this section we look at the evidence that there is significant synonymous rate heterogeneity among sites in the chloroplast genome as a function of context, and that the variation in substitution bias presented previously results from differences between the rate of transitions and the rate of transversions in the different contexts.

The relationship between context and synonymous substitution rate in cpDNA was examined using four different data sets; noncoding regions from rice and maize, fourfold degenerate sites from 19 protein-coding genes from the same two species, and fourfold degenerate sites from multiple alignments of both *rbcL* and *ndhF*, using those sites with conserved contexts as described previously (Morton, 1997b). For the pairwise analyses, silent substitution rate was calculated separately for sites in each AT context and for the multiple alignments, the average number of inferred changes per site for each of the three possible AT contexts was calculated. For all four data sets the substitution rate of each context was normalized by the context with the highest rate for that data set (Fig. 6). In every case the rate of substitution increases with decreasing AT content of the two

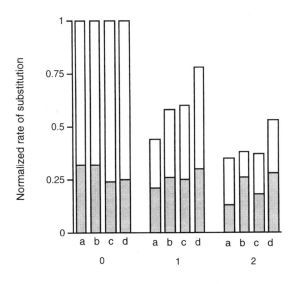

A+T content of the flanking bases

FIG. 6. Rates of transitions (open) and transversions (shaded) for the four analyses described in the text; *rbcL* (a), *ndhF* (b), coding sequences of rice and maize (c), and noncoding sequences of rice and maize (d). For each analysis, the average rate was determined separately for sites with the three different possible AT contents of the two flanking bases and was then normalized by the 0 flanking AT context, which had the highest average rate in each analysis.

flanking bases. The average relative rate of substitution in AT-rich contexts is roughly one third the rate in GC-rich contexts and the results are significant at the 1% level for every data set. In addition, the different types of substitution vary among the contexts in different manners. The absolute rate of transition substitutions varies much more widely among contexts than the rate of transversion (see Fig. 6) giving rise to the variation in substitution bias described in the previous section.

An analysis of the rate of transitions at twofold degenerate sites shows the same correlation with context. Looking at *rbcL*, if we consider only codon groups with an A or a T at the second position, which is 7 of the 10 twofold degenerate codon groups, those sites with a G or a C at the 3' neighboring position average 13.3 inferred transitions whereas those with a 3' A or T average only 8.3 inferred transitions. For the same comparison in rice and in maize coding sequences, the synonymous substitution rates in the two sets of sites are 0.065 and 0.042 respectively, a difference that is significant at the 5% level. Therefore, neighboring base composition influences the rate of silent substitution at twofold degenerate sites as well as at fourfold degenerate sites.

In mammalian pseudogenes, CpG to TpG transitions occur at an increased rate, probably due to spontaneous deamination (Bulmer, 1986). Although this could potentially explain the previous observations, the variation in transition rate among contexts in cpDNA does not appear to be the result of a similar mechanism. The contexts GXG, GXC, CXG, and CXC all have high rates of substitution, and the context GXC, which can have no CpG site on either strand, has a rate equal to the rate at CXG sites in both coding and in noncoding sequences (data not shown). In addition, a comparison of closely related *rbcL* sequences with an outgroup reference sequence shows that all transitions, regardless of whether they are in the TA → CG or CG → TA direction, increase in frequency in a higher GC context (Morton, unpublished data).

Distribution of Synonymous Substitutions in *rbcL*

One result of the strong influence of context on substitution rate is that the variation in average substitution rate among fourfold degenerate sites in *rbcL* among contexts can be divided into three separate distributions as a function of context (Fig. 7). It appears, however, that although flanking base AT content explains some proportion of the variation among fourfold degenerate sites in *rbcL*, significant variation still exists within each distribution. There are indications that additional context features might account for the remaining variation. Comparing monocot to dicot sequences in terms of the number of inferred substitutions at each fourfold degenerate site shows that sites in the distributions of both Fig. 7a and Fig. 7b are strongly correlated ($r = 0.62$ and $r = 0.56$, respectively). Therefore, even when flanking AT content is accounted for, variation among sites is similar in different angiosperm lineages. (The sites in the distribution in Fig. 7c were not compared because there are only 15.) This correlation suggests that there are site-specific features, perhaps local base composition or secondary structure (e.g., Todd and Glickman, 1982), that are responsible for at least some of the variation observed in the different distributions in Fig. 7. It is also possible that changes at different fourfold degenerate sites are correlated, or are correlated with amino acid substitutions. All of these possibilities will have to be examined if we are to explain the variation in substitution rate among fourfold degenerate sites in the flowering plants.

There is some direct evidence that additional context features may be associated with synonymous substitution rate variation at fourfold degenerate sites of *rbcL* in the flowering plants. When only sites with an AT content of flanking base of zero are considered (Fig. 7a), the presence of a 5′ CC on either strand is associated with increases in substitution rate. Sites with either a 5′ CC or a 3′ GG have significantly higher rates of substitution and the three sites in *rbcL* that have the context CCXGG, where X indicates the third codon position, are all extremely variable (Table XIII). The effect of a 5′ CC is also made apparent by the fact that, of the 10 most variable fourfold degenerate sites in *rbcL*, 7 have a 5′ CC (3 of which are the CCXGG sites that have one on each strand) whereas only 1 of the 10 most strongly conserved fourfold degenerate sites has a 5′ CC (data not shown). In addition, the effect of a 5′ CC is not limited to fourfold degenerate sites; 4 of the 5 most variable twofold degenerate sites have a 3′ GG, which is a significant association at the 5% level. The identification of a

TABLE XIII. Association between the Presence of a 5′ CC Dinucleotide and Average Substitution Rate at Fourfold Degenerate Sites of *rbcL*[a]

Context[b]	Sites	Average rate
GGX	45	22.1
CCX	18	39.6
XCC	6	19.7
XGG	20	32.3
CCXGG	3	67.0

[a] For those sites with 2 flanking G or C nucleotides.
[b] X represents the site of substitution.

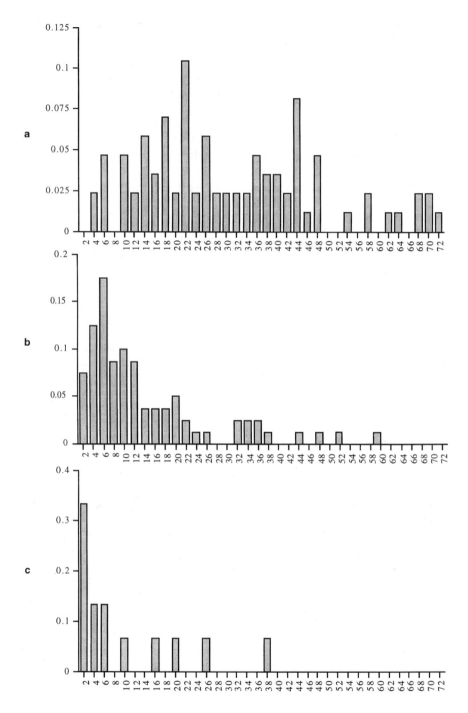

Substitutions per site

specific dinucleotide that is associated with rate heterogeneity suggests that it might be possible to explain a great deal of synonymous rate variation among sites in cpDNA as a function of context, but this needs to be examined further.

Context and Synonymous Substitution Rate of Chloroplast Genes

One of the interesting implications of the variation in synonymous substitution rate as a function of local context is that there is a source of significant variation in the mutation rate between genes such that, even in the absence of selection, we can expect to observe significant variation in the synonymous substitution rate. Genes rich in amino acids with GC-rich codons, such as glycine, proline, and arginine, should have higher rates of substitution than those rich in amino acids such as lysine and isoleucine that are coded by AT-rich codons due to differences in average context between the two genes.

This potential effect can be tested by calculating an average context for every gene, defined here as the AT Index (ATI) that is given in Equation (1) where A is the AT content of the two neighbors of each third position and C is the number of codons.

$$ATI = \sum A^2/C. \tag{1}$$

When the average ATI for homologous genes from rice and maize is plotted against Ks as calculated by the method of Li (1993) the expected inverse correlation is observed (Fig. 8). This source of variation in mutation rate among loci, and potentially at a single locus over time, has important implications for our understanding of the molecular evolution of cpDNA and our interpretation of synonymous rate variation among loci. It also indicates that an initial assessment of which locus might have a sufficient level variation for a particular phylogenetic analysis could be enhanced by considering average contexts. Genes with a high use of GC-rich codons are likely to prove more appropriate for closely related taxa and AT-rich genes are likely to be more appropriate for more distantly related taxa.

FIG. 7. The distribution of inferred substitutions per fourfold degenerate site plotted for sites with (a) two flanking G or C nucleotides, (b) one flanking G or C, and (c) two flanking A or T nucleotides.

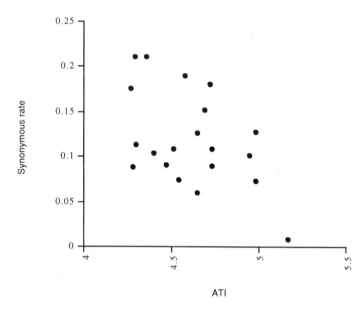

FIG. 8. Plot comparing Ks (Li, 1993) to the ATI (AT Index, see text) for 20 protein-coding genes from rice and maize.

Context Dependency in Lineages Other than the Angiosperms

All of the analyses of flowering plant cpDNA sequences presented previously demonstrate a process of substitution that is a complex function of local context. As has been discussed, it seems likely that it occurs at the level of mutation. If this is the case then what do we expect to observe in other genomes and in other lineages? Is the effect primarily a function of the structure of DNA, the replication and repair machinery, or do both play a role?

To test for context dependency in other plastid lineages, four *rbcL* data sets were used. The first consisted of red algae representing the Cerinales (Freshwater *et al.*, 1994), the second of green algae from the Volvocaceae (Nozaki *et al.*, 1997), the third of ferns (Hasebe *et al.*, 1994) and the last of members of the pine family (unpublished sequences in GenBank). For each data set substitutions at fourfold degenerate sites were analyzed as described in Morton (1997b). Parsimony trees generated from an equal weighting heuristic search were used to generate a 50% majority-rule consensus tree for each analysis.

The comparisons of substitution bias and rate to flanking base AT content are presented in Table XIV. The results indicate that the pattern of

TABLE XIV. Context Dependency of Substitutions at Fourfold Degenerate Sites of the *rbcL* Locus from Various Plastid Lineages

	Green algae		Red algae		Ferns		*Pinus*	
AT^a	$\dfrac{Tv}{Tv+Ts}$	$Rate^b$	$\dfrac{Tv}{Tv+Ts}$	Rate	$\dfrac{Tv}{Tv+Ts}$	Rate	$\dfrac{Tv}{Tv+Ts}$	Rate
0	76.1	3.3	74.3	5.1	23.1	11.6	14.6	0.48
1	86.6	3.1	73.0	5.1	27.5	8.6	26.1	0.22
2	97.2	3.3	77.1	5.0	42.3	10.5	57.1	0.32
χ^2	$p < 0.05$	NS	NS	NS	$p < 0.01$	$p < 0.01$	$p < 0.05$	$p < 0.05$

^{a} AT content of the two nucleotides immediately flanking the substitution.
^{b} Inferred number of substitutions per site.

context dependency observed in flowering plants is not present in all plastid lineages. The red algae show absolutely no variation in either feature in different contexts whereas the green algae show a small amount of variation in substitution bias but not in rate. In both cases, however, the bias to transversions is very high, over 73% of all substitutions in every context. Therefore, the substitution dynamics are dramatically different than in the flowering plants.

In the fern and the pine data sets, the situation more closely resembles what is observed in the flowering plants in that bias toward transversions increases with increasing AT context. The variation in rate, however, is more difficult to interpret. There is significant variation in each case but the extreme AT contexts show relatively high rates, unlike in the angiosperm where a consistent trend was observed. In addition, the overall proportion of substitutions that are transversions is much more similar to the bias observed in flowering plants than in the algae.

The results in Table XIV make it clear that it is difficult to generalize about context dependency of substitutions. There is a great deal of variation in substitution dynamics among lineages as well as differences in the relationship between context and substitution dynamics. One interesting feature is the difference between algae and flowering plant lineages in terms of overall substitution bias. In Table XV the number of occurrences of each transversion type, without a consideration for direction, is shown for the algae and for the flowering plant *rbcL* genes. What is apparent is that the algae are essentially alternating between two states, A and T, something that is not nearly as pronounced in the angiosperm. Given the high AT content of the algae (Table I), it will be interesting to investigate how these two features are related. Because the algae are under fairly stringent selection for codon usage, especially at the *rbcL* locus (Table V, discussed previously), the

TABLE XV. Number of the Different Types of Transversion Substitution Observed in Different Lineages

Taxa	A ↔ T	G ↔ T	A ↔ C	G ↔ C
Rhodophytes	1,116	95	59	5
Chlorophytes	382	46	4	0
Angiosperms	283	448	295	183

issue of whether the substitution dynamics of the algae reflect selection for codon usage or possibly selection for increased AT bias should be pursued. In particular, it would be useful to analyze substitutions in noncoding regions of algae DNA for comparison to fourfold degenerate sites of *rbcL*.

CONCLUSIONS

The plastid genome has features that make it a very interesting model for future studies in molecular evolution. The pattern of evolution at the *psbA* locus, and to a lesser degree the *rbcL* locus, of flowering plants suggests possible recent shifts in selective pressure on codon usage and, perhaps, on genome-wide processes, a hypothesis that awaits further testing. In addition, the drastic differences in substitution dynamics between algae and plants suggest either variation in the replication/repair processes or differences in selection. If the latter is the case, merging the two lines of research may prove insightful. In particular, the high rate of A <–> T substitutions in algae, coupled to the observation of strong discriminating selection on codon usage, may prove an excellent system for testing the role of selection on genome composition bias. Finally, the significant rate variation among fourfold degenerate sites in angiosperm cpDNA offers an excellent opportunity to advance our knowledge of how various factors affect substitution dynamics and how a violation of the assumption that sites are independent influences analyses of molecular evolution and phylogenetic reconstruction.

ACKNOWLEDGMENTS

I would like to thank several people for discussion of this work in the past, particularly Brandon S. Gaut, Gerald H. Learn Jr., Michael T. Clegg,

Richard Morton, Brian Golding, and Walter Fitch. Jeff Thorne deserves special thanks for very generously giving up some of his time to perform the maximum likelihood alignments. This work was supported in part by grant MCB-9727906 from NSF.

REFERENCES

Akashi, H., 1995, Inferring weak selection from patterns of polymorphism and divergence at "silent" sites in Drosophila DNA, *Genetics* **139:**1067–1076.

Albert, V. A., Backlund, A., Bremer, K., Chase, M. W., Manhart, J. R., Mishler, B. D., and Nixon, K. C., 1994, Functional constraints and rbcL evidence for land plant phylogeny. *Annals of the Missouri Botanical Garden* **81:**534–567.

Andersson, S. G. E., and Kurland, C. G., 1990, Codon preferences in free-living microorganisms, *Microbiological Reviews* **54:**198–210.

Andersson, S. G. E., and Sharp, P. M., 1996, Codon usage in the *Mycobacterium tuberculosis* complex, *Microbiology* **142:**915–925.

Berg, O. G., and Martelius, M., 1995, Synonymous substitution-rate constants in *Escherichia coli* and *Salmonella typhimurium* and their relationship to gene expression and selection pressure, *J. Mol. Evol.* **41:**449–456.

Blake, R. D., Hess, S. T., and Nicholson-Tuell, J., 1992, The influence of nearest neighbors on the rate and pattern of spontaneous point mutations, *J. Mol. Evol.* **34:**189–200.

Brown, W. M., Prager, E. M., Wang, A., and Wilson, A. C., 1982, Mitochondrial DNA sequences of primates: Tempo and mode of evolution, *J. Mol. Evol.* **18:**225–239.

Bulmer, M., 1986, Neighboring base effects on substitution rates in pseudogenes, *Mol. Biol. Evol.* **3:**322–329.

Chandrasekharan, U. M., Sanker, S., Glynias, M. J., Karnik, S. S., and Husain, A., 1996, Angiotensin II-forming activity in a reconstructed ancestral chymase, *Science* **271:**502–505.

Chase, M. W., Soltis, D. E., Olmstead, R. G., *et al.*, 1993, Phylogenetics of seed plants: An analysis of nucleotide sequences from the plastid gene *rbcL*, *Annals of the Missouri Botanical Gardens* **80:**528–580.

Clegg, M. T., 1993, Chloroplast gene sequences and the study of plant evolution, *Proc. Natl. Acad. Sci. USA* **90:**363–367.

Clegg M. T., Ritland, K., and Zurawski, G., 1986, Processes of chloroplast DNA evolution, in: *Evolutionary Processes and Theory* (S. Karlin and E. Nevo, Eds.). pp. 275–294, Academic Press, New York.

Curtis, S. E., and Clegg, M. T., 1984, Molecular evolution of chloroplast DNA sequences, *Mol. Biol. Evol.* **1:**291–304.

Delwiche, C. F., and Palmer, J. D., 1996, Rampant horizontal transfer and duplication of rubisco genes in eubacteria and plastids, *Mol. Biol. Evol.* **13:**873–882.

Downie, S. R., and Palmer, J. D., 1992, Use of chloroplast DNA rearrangements in reconstructing plant phylogeny, in: *Molecular Systematics of Plants* (P. S. Soltis, D. E. Soltis, and J. J. Doyle, Eds.), pp. 14–35, Chapman and Hall, New York, New York.

Eyre-Walker, A., and Bulmer, M., 1995, Synonymous substitution rates in enterobacteria, *Genetics* **140:**1407–1412.

Eyre-Walker, A., and Gaut, B. S., 1997, Correlated substitution rates among plant genomes, *Mol. Biol. Evol.* **14:**455–460.

Felsenstein, J., 1993, *PHYLIP 3.5-phylogeny inference package, version 3.5*, distributed by the author, Department of Genetics, University of Washington, Seattle.

Felsenstein, J., and Churchill, G. A., 1996, A hidden Markov model approach to variation among sites in the rate of evolution, *Mol. Biol. Evol.* **13**:93–104.

Fitch, W. M., 1971, Toward defining the course of evolution: Minimum change for a specific tree topology, *Syst. Zoology* **20**:406–416.

Freshwater D. W., Fredericq, S., Butler, B. S., and Hommersand, M. H., 1994, A gene phylogeny of the red algae (Rhodophyta) based on plastid *rbcL*, *Proc. Natl. Acad. Sci. USA* **91**:7281–7285.

Gaut, B. S., Morton, B. R., McCaig, B. C., and Clegg, M. T., 1996, Substitution rate comparisons between grasses and palms: Synonymous rate differences at the nuclear gene *Adh* parallel rate differences at the plastid gene *rbcL*, *Proc. Natl. Acad. Sci. USA* **93**:10274–10279.

Gaut, B. S., Muse, S. V., Clark, W. D., and Clegg, M. T., 1992, Relative rates of nucleotide substitution at the *rbcL* locus of monocotyledenous plants, *J. Mol. Evol.* **35**:292–303.

Gojobori, T., Li, W.-H., and Graur, D., 1982, Patterns of nucleotide substitution in pseudogenes, *J. Mol. Evol.* **18**:360–369.

Grosjean, H., and Fiers, W., 1982, Preferential codon usage in prokaryotic genes: The optimal codon–anticodon interaction energy and the selective codon usage in efficiently expressed genes, *Gene* **18**:199–209.

Grosjean, H., Houssier, C., and Cedergren, R., 1986, Anticodon–Anticodon interaction and tRNA sequence comparison: Approaches to codon recognition, in: *Structure and Dynamics of RNA* (P. H. van Knippenberg and C. W. Hilbers, Eds.), pp. 161–174, Plenum Press, New York and London.

Gu, X., and Zhang, J., 1997, A simple method for estimating the parameter of substitution rate variation among sites, *Mol. Biol. Evol.* **14**:1106–1113.

Gu, X., Fu, Y. X., and Li, W. H., 1995, Maximum likelihood estimation of the heterogeneity of substitution rate among nucleotide sites, *Mol. Biol. Evol.* **12**:546–557.

Hallick, R. B., and Bairoch, A., 1994, Proposals for the naming of chloroplast genes. III. Nomenclature for open reading frames encoded in chloroplast genomes, *Plant Mol. Biol. Reptr.* **12**:S29–30.

Hallick, R. B., Hong, L., Drager, R. G., Favreau, M. R., Montfort, A., Orsat, B., Spielman, A., and Stutz, E., 1993, Complete sequence of *Euglena gracilis* chloroplast DNA, *Nuc. Acids Res.* **21**:3537–3544.

Hartl, D. L., Moriyama, E. N., and Sawyer, S., 1994, Selection intensity on codon bias, *Genetics* **138**:227–234.

Hasebe, M., Omori, T., Nakazawa, M., Sano, T., Kato, M., and Iwatsuki, K., 1994, *rbcL* gene sequences provide evidence for the evolutionary lineages of leptosporangiate ferns, *Proc. Natl. Acad. Sci. USA* **91**:5730–5734.

Hess, S. T., Blake, J. D., and Blake, R. D., 1994, Wide variations in neighbor-dependent substitution rates, *J. Mol. Biol.* **236**:1022–1033.

Hillis, D. M., Huelsenbeck, J. P., and Cunningham, C. W., 1994, Application and accuracy of molecular phylogenies, *Science* **264**:671–677.

Huelsenbeck, J., and Hillis, D. M., 1993, Success of phylogenetic methods in the four-taxon case, *Syst. Biol.* **42**:247–264.

Ikemura, T., 1985, Codon usage and tRNA content in unicellular and multicellular organisms, *Mol. Biol. Evol.* **2**:13–35.

Jermann, T. M., Opitz, J. G., Stackhouse, J., and Benner, S. A., 1995, Reconstructing the evolutionary history of the artidactyl ribonuclease superfamily, *Nature* **374**:57–59.

Jin, L., and Nei, M., 1990, Limitations of the evolutionary parsimony method of phylogenetic analysis, *Mol. Biol. Evol.* **7**:82–102.

Jones M., Wagner, R., and Radman, M., 1987, Repair of a mismatch is influenced by the base composition of the surrounding nucleotide sequence, *Genetics* **115**:605–610.

Kellogg, E. A., and Juliano, N. D., 1997, The structure and function of RuBisCO and their implications for systematic studies, *Amer. Journal of Bot.* **84**:413–428.

Kelly, C., 1994, A test of the Markovian model of DNA evolution, *Biometrics* **50**:653–664.

Kimura, M., 1980, A simple method for estimating evolutionary rate of base substitutions through comparative studies of nucleotide sequences, *J. Mol. Evol.* **16**:111–120.

Klein, R. R., and Mullet, J. E., 1986, Regulation of chloroplast-encoded chlorophyll-binding protein translation during higher plant chloroplast biogenesis, *J. Biol. Chem.* **261**:11138–11145.

Kowallik, K. V., Stoebe, B., Schaffran, I., and Freier, U., 1995, The chloroplast genome of a chlorophyll a+c-containing alga, *Odontella sinensis*, *Plant Mol. Biol. Reptr.* **13**:336–342.

Lewontin, R. C., 1989, Inferring the number of evolutionary events from DNA coding sequence differences, *Mol. Biol. Evol.* **6**:15–32.

Li, W.-H., 1987, Models of nearly neutral mutations with particular implications for nonrandom usage of synonymous codons, *J. Mol. Evol.* **24**:337–345.

Li, W.-H., 1993, Unbiased estimation of the rates of synonymous and nonsynonymous substitution, *J. Mol. Evol.* **36**:96–99.

Li, W.-H., Wu, C. I., and Luo, C.-C., 1984, Nonrandomness of point mutation as reflected in nucleotide substitutions in pseudogenes and its evolutionary implications, *J. Mol. Evol.* **21**:58–71.

Lobry, J. R., and Gautier, C., 1994, Hydrophobicity, expressivity and aromaticity are the major trends of amino-acid usage in 999 *Escherichia coli* chrmosome-encoded genes, *Nuc. Acids Res.* **22**:3174–3180.

Lockhart, P. J., Penny, D., Hendy, M. D., Howe, C. J., Beanland, T. J., and Larkum, A. W. D., 1992, Controversy on chloroplast origins, *FEBS Letters* **301**:127–131.

Long, M., and Gillespie, J. H., 1991, Codon usage divergence of homologous vertebrate genes and codon usage clock, *J. Mol. Evol.* **32**:6–15.

Malcolm, B. A., Wilson, K. P., Matthews, B. W., Kirch, J. F., and Wilson, A. C., 1990, Ancestral lysozymes reconstructed, neutrality tested, and thermostability linked to hydrocarbon packing, *Nature* **345**:86–89.

Mendelman, L. V., Boosalis, M. S., Petruska, J., and Goodman, M. F., 1989, Nearest neighbor influences on DNA polymerase insertion fidelity, *J. Biol. Chem.* **264**:14415–14423.

Mindell, D. P., and Thacker, C. E., 1996, Rates of molecular evolution: Phylogenetic issues and applications, *Ann. Rev. Ecol. Syst.* **27**:279–303.

Morton, B. R., 1993, Chloroplast DNA codon use: Evidence for selection at the *psbA* locus based on tRNA availability, *J. Mol. Evol.* **37**:273–280.

Morton, B. R., 1994, Codon use and the rate of divergence of land plant chloroplast genes, *Mol. Biol. Evol.* **11**:231–238.

Morton, B. R., 1995, Neighboring base composition and transversion/transition bias in a comparison of rice and maize chloroplast noncoding regions, *Proc. Natl. Acad. Sci. USA* **92**:9717–9721.

Morton, B. R., 1996, Selection on the codon bias of *Chlamydomonas reinhardtii* chloroplast genes and the plant *psbA* gene, *J. Mol. Evol.* **43**:28–31.

Morton, B. R., 1997a, Rates of synonymous substitution do not indicate selective constraints on the codon use of the plant *psbA* gene, *Mol. Biol. Evol.* **14**:412–419.

Morton, B. R., 1997b, The influence of neighboring base composition on substitutions in plant chloroplast coding sequences, *Mol. Biol. Evol.* **14**:189–194.

Morton, B. R., 1998, Selection on the codon bias of chloroplast and cyanelle genes in different plant and algal lineages, *J. Mol. Evol.* **46**:449–459.

Morton, B. R., and Clegg, M. T., 1995, Neighboring base composition is strongly correlated with base substitution bias in a region of the chloroplast genome, *J. Mol. Evol.* **41:**597–603.

Morton, B. R., and Levin, J. A., 1997, The atypical codon use of the plant *psbA* gene may be the remnant of an ancestral bias, *Proc. Natl. Acad. Sci. USA* **94:**11434–11438.

Morton, B. R., Oberholzer, V. M., and Clegg, M. T., 1997, The influence of specific neighboring bases on substitution dynamics in noncoding regions of the plant chloroplast genome, *J. Mol. Evol.* **45:**227–231.

Mullet, J. E., and Klein, R. R., 1987, Transcription and RNA stability are important determinants of higher plant chloroplast RNA levels, *EMBO. J.* **6:**1571–1579.

Muse, S. V., 1995, Evolutionary analyses of DNA sequences subject to constraints on secondary structure, *Genetics* **139:**1429–1439.

Muse, S. V., and Gaut, B. S., 1997, Comparing patterns of nucleotide substitution rates among chloroplast loci using the relative ratio test, *Genetics* **146:**393–399.

Nei, M., 1987, *Molecular Evolutionary Genetics.* Columbia University Press, New York.

Nozaki, H., Ito, M., Sano, R., and Uchida, H., 1997, Phylogenetic analysis of Yamagishiella and Platydorina (Volvocaceae, Chlorophyta) based on *rbcL* gene sequences, *J. Phycology* **33:**272–278.

Palmer, J. D., 1991, Plastid chromosomes: Structure and evolution, in: *The Molecular Biology of Plastids* L. Bogorad and I. K. Vasil., Eds. (pp. 5–53) Vol. 7 of *Cell Culture and Somatic Cell Genetics in Plants* (I. K. Vasil, Ed.) Academic Press, San Diego.

Palmer, J. D., 1993, A genetic rainbow of plastids, *Nature* **364:**762–763.

Palmer, J. D., 1997, Organelle genomes: Going going gone! *Science* **275:**790–791.

Palmer, J. D., and Delwiche, C. F., 1996, Second-hand chloroplasts and the case of the disappearing nucleus, *Proc. Natl. Acad. Sci. USA* **93:**7432–7435.

Petruska, J., and Goodman, M. F., 1985, Influence of neighboring bases on DNA polymerase insertion and proofreading fidelity, *J. Biol. Chem.* **260:**7533–7539.

Radman, M., and Wagner, R., 1986, Mismatch repair in *Escherichia coli, Ann. Rev. Genet.* **20:**523–538.

Reith, M., 1995, Molecular biology of Rhodophyte and Chromophyte plastids, *Ann. Rev. Plant Physiol. Plant Mol. Biol.* **46:**549–575.

Ritland, K., and Clegg, M. T., 1987, Evolutionary analysis of plant DNA sequences, *American Naturalist* **130:**S74–S100.

Saitou, N., and Nei, M., 1987, The neighbor-joining method: A new method for reconstructing phylogenetic trees, *Mol. Biol. Evol.* **4:**406–425.

Savolainen, V., Spichiger, R., and Manen, J.-F., 1997, Polyphyletism of Cestrales deduced from a chloroplast noncoding DNA region, *Mol. Phylogenet. Evol.* **7:**145–157.

Sharp, P. M., 1991, Determinants of DNA sequence divergence between *Escherichia coli* and *Salmonella typhimurium*: Codon usage, map position, and concerted evolution, *J. Mol. Evol.* **33:**23–33.

Sharp, P. M., and Li, W.-H., 1987b, The rate of synonymous substitution in enterobacterial genes is inversely related to codon usage bias, *Mol. Biol. Evol.* **4:**222–230.

Sharp, P. M., and Li, W.-H., 1987a. The codon adaptation index—A measure of directional synonymous codon usage bias, and its potential applications, *Nuc. Acids Res.* **15:**1281–1295.

Sharp, P. M., Tuohy, T. M. F., and Mosurski, K. R., 1986, Codon usage in yeast: Cluster analysis clearly differentiates highly and lowly expressed genes, *Nuc. Acids Res.* **14:**5125–5139.

Sharp, P. M., Cowe, E., Higgins, D. G., Shields, D. C., Wolfe, K. H., and Wright, F., 1988, Codon usage patterns in *Escherichia coli, Bacillus subtilis, Saccharomyces cerevisiae, Schizosaccharomyces pombe, Drosophila melanogaster* and *Homo sapiens*; A review of the considerable within-species diversity, *Nuc. Acids Res.* **16:**8207–8211.

Shields, D. C., Sharp, P. M., Higgins, D. G., and Wright, F., 1988, "Silent" sites in *Drosophila* genes are not neutral: Evidence of selection among synonymous codons, *Mol. Biol. Evol.* **5:**704–716.

Stewart, C.-B., 1995, Active ancestral molecules, *Nature* **374:**12–13.

Swofford, D. L., and Olsen, G. J., 1990, Phylogeny Reconstruction, in: *Molecular Systematics* (D. M. Hillis and C. Moritz, Eds.), pp 411–501, Sinauer Associates, Sunderland.

Tateno, Y., Takezaki, N., and Nei, M., 1994, Relative efficiencies of the maximum-likelihood, neighbor-joining and maximum parsimony methods when substitution rate varies with site, *Mol. Biol. Evol.* **11:**261–277.

Thorne, J. L., and Churchill, G. A., 1995, Estimation and reliability of molecular sequence alignments, *Biometrics* **51:**100–113.

Todd, P. A., and Glickman, B. W., 1982, Mutational specificity of UV light in *Escherichia coli*: Indications for a role of DNA secondary structure, *Proc. Natl. Acad. Sci. USA* **79:**4123–4127.

Topal, M. D., and Fresco, J. R., 1976, Complementary base pairing and the origin of substitution mutations, *Nature* **263:**285–289.

van den Hoek, C, 1995, *Algae: An Introduction to Phycology*, Cambridge University Press, Cambridge, New York.

Wakeley, J., 1994, Substitution-rate variation among sites and the estimation of transition bias, *Mol. Biol. Evol.* **11:**436–442.

Wolfe, K. H., Li, W. H., and Sharp, P. M., 1987, Rates of nucleotide substitution vary greatly among plant mitochondrial, chloroplast, and nuclear DNAs. *Proc. Natl. Acad. Sci. USA* **84:**9054–9058.

Wolfe, K. H., Morden, C. W., Ems, S. C., and Palmer, J. D., 1992, Rapid evolution of the plastid translational apparatus in a nonphotosynthetic plant: Loss or accelerated sequence evolution of tRNA and ribosomal protein genes, *J. Mol. Evol.* **35:**304–317.

Yang, Z., 1994, Maximum likelihood estimation of phylogeny from DNA sequences when substitution rates differ over sites: Approximate methods, *J. Mol. Evol.* **39:**306–314.

Yang, Z., 1996, Among-site rate variation and its impact on phylogenetic analysis, *Trends Ecol. Evol.* **11:**367–372.

Yang, Z., and Kumar, S., 1996, Approximate methods for estimating the pattern of nucleotide substitution and the variation of substitution rates among sites, *Mol. Biol. Evol.* **13:**650–659.

Yang, Z. H., Kumar, S., and Nei, M., 1995, A new method of inference of ancestral nucleotide and amino acid sequences, *Genetics* **141:**1641–1650.

Zhang, J. Z., and Nei, M., 1997, Accuracies of ancestral amino acid sequences inferred by the parsimony, likelihood, and distance methods, *J. Mol. Evol.* **44:**S139–S146.

Zurawski, G., and Clegg, M. T., 1987, Evolution of higher plant chloroplast DNA-encoded genes: Implications for structure–function and phylogenetic studies, *Ann. Rev. Plant Physiol.* **38:**391–418.

3

The Origin of the Mineral Skeleton in Chordates

JERZY DZIK

INTRODUCTION

Most of the evolution of the vertebrate mineralized dermal skeleton, which has resulted in its present complexity, is traceable within the crossopterygian–tetrapod lineage. In acanthodianlike ancestors of these higher vertebrates the dermal skeleton was restricted to scales, locally coalescing into dermal plates or modified into oral teeth. The basic unit of such primitive vertebrate skeleton is the odontode, a denticle forming at the tip of an ectomesenchymal dental papilla under its ectodermal epithelium (recently reviewed by Smith and Hall, 1993; Smith, 1995). In most fishes and extinct armored agnathans the external layer of mineralized dermal tissue is enameloid, developing as a result of mineralization of a matrix produced by both ectodermal and ectomesenchymal cells of the dental papilla, or histogenetically similar complex tissue. Below this layer, purely ectomesenchymal in origin derivatives of dentine develop. However, in the most primitive actinopterygian fishes, ganoine, an enamel homolog (Sire *et al.*, 1987) develops, instead of enameloid. Odontodes composed of acellular bone (aspidin) and capped with enamel characterize the Ordovician agnathan *Eriptychius* (Smith and Hall, 1990; M. P. Smith *et al.*, 1995; M. M. Smith *et al.*, 1996). The original odontode organization was thus, as can be judged on the basis of this pattern of distribution of ectodermally secreted mineral tissues, a thick and dense cap built of a mineral laminated tissue, ectodermally secreted

JERZY DZIK • Institute of Paleobiology PAN, 00-818 Warsaw, Poland.

Evolutionary Biology, Volume 31, edited by Max K. Hecht *et al.* Kluwer Academic / Plenum Publishers, New York, 2000.

from outside by ameloblasts (Smith, 1995), and less compact basal tissue filling the internal denticle cavity, secreted in a rather irregular way by ectomesenchymally derived odontoblasts, which may or may not be incorporated into the mineral tissue (Smith and Hall, 1990). Such a structure characterizes elements of the oral apparatuses of the conodonts, extinct organisms preceding in their origin the armored agnathans. The conodonts are the oldest and the most primitive chordates bearing a well-developed mineralized dermal skeleton.

The mode of formation of the phosphatic skeleton, which characterizes both the conodont elements and the vertebrate dermal scales, is unique for these two kinds of sclerites and is unknown in any other organisms. Their homology seems to be based on strong evidence (Smith, 1995). It is highly unlikely that this could have evolved independently twice in metazoan phylogeny. Rather, the genetic and developmental patterns that evolved at the earliest stages of phylogeny of the oral apparatus of the common ancestor of the conodonts and agnathans were applied by the latter to the secretion of dermal scales (M. M. Smith et al., 1996; M. P. Smith et al., 1996). This could have been connected with a change from the pelagic to near-bottom mode of life and to a different feeding style, when the weight of newly constructed armor was no longer an obstacle to building a mineral protective armor and there was also no longer a need for a predatory grasping apparatus.

This chapter contains a review of the paleontological evidence available on the earliest evolution of the chordate mineral skeleton and its correspondence to soft anatomy. The data on which the reviewed anatomical and phylogenetic inferences are based are of varied nature, ranging from isolated mineral denticles to complete organisms with details of soft organs preserved. This richness of the fossil record is a rather unexpected result of paleontological research since the late 1980s. It is now possible to merge the information provided by statistical studies on isolated skeletal elements of conodonts (reviewed by Dzik, 1991a), articulated skeletal parts (Aldridge et al., 1987, 1995), and complete imprints of the soft body (Briggs et al., 1983; Aldridge et al., 1986, 1993). The Cambrian soft-bodied faunas continue to supply new data on even more primitive chordates (e.g., Dzik, 1995); the Ordovician is also important in discussion of the phylogeny of vertebrates (e.g., Sansom et al., 1996). The available evidence is incomplete, typical of all paleontological material, but is unrivalled because it comes directly from those geological epochs in which the evolutionary diversification actually took place, or at least from a time closely following establishment of the main clades. The fossil record, therefore, regularly provides challenging tests to the relationships between the Recent chordates, established on neontological data (see Forey and Janvier, 1993).

The question of the origin of the mineral skeleton in chordates will start from presentation of the basic data on the organization and evolution of the skeletal structures in the conodont body, their oral apparatuses. An interesting feature of conodont elements is the occurrence of clear imprints of secretory cells on their oral surface. The imprints provide insight, at the cellular level, into developmental processes in these extinct chordates, many of which are up to 500 million years old. An attempt to restore the organization of the secretory organ and mechanisms of morphogenesis of particular morphological structures on the basis of cell imprint distribution is presented. The oral grasping apparatuses of chordates, as well as their elaborate sensory head organs, appear to be even more ancient in origin than those of true conodonts and are present in the Middle Cambrian probable "paraconodont" *Odontogriphus* (Conway Morris, 1976*a*), and perhaps even in the oldest known chordate *Yunnanozoon*, of Early Cambrian age (Dzik, 1995; Shu *et al.*, 1996*a* interpreted another fossil from the same locality as a chordate). These organisms differ strongly from conodonts and more advanced chordates in their anatomy, as well as from the generally assumed ancestral chordate Bauplan (but see Lacalli, 1996; Williams and Holland, 1996). A scenario of possible evolutionary transformations that may explain this deviation from expectations, consistent with the fossil evidence although radically unorthodox from a zoological perspective, is proposed. According to this model, the evolutionary origin of the oral apparatus with its mineralized skeleton took place well within the invertebrate grade of the sequence of events leading to the chordates.

ARCHITECTURE OF CONODONT APPARATUSES

The chordate affinity of conodonts is based on the occurrence of V-shaped muscle blocks and an axial notochord, an asymmetric caudal fin, and paired eyes (Fig. 1*a*; Briggs *et al.*, 1983; Dzik, 1986; Aldridge *et al.*, 1986, 1993; proposed to be otic capsules by Pridmore *et al.*, 1997), as well as the probable homology of their mineralized tissues with those of the vertebrates (Schmidt and Müller, 1964; Dzik, 1976, 1986; Smith, 1995). The conodonts are classified in several orders (Sweet, 1988; Dzik, 1991*b*), most of which are distinguished on the basis of the organization of their oral apparatuses. Soft tissue anatomy is also relatively well known, owing to imprints of representatives of the two most diverse orders. *Clydagnathus* from the early Carboniferous of Scotland (Briggs *et al.*, 1983; Aldridge *et al.*, 1986, 1993) is representative of the Ozarkodinida, whereas *Promissum*, from the late Ordovician of South Africa (Aldridge and Theron, 1993; Gabbott *et al.*,

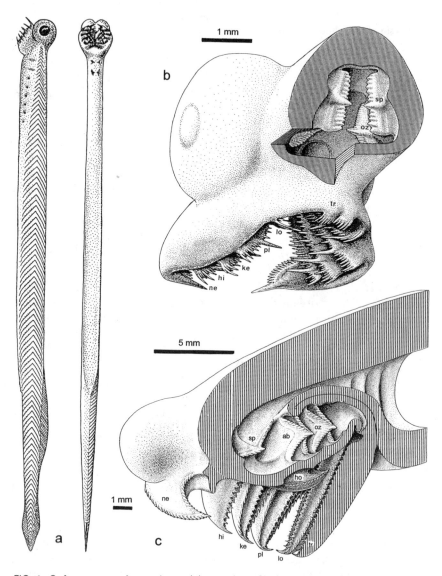

FIG. 1. Soft anatomy of conodonts: (a) conodont *Clydagnathus*? cf. *cavusformis* from the Early Carboniferous (Viséan) Granton Shrimp Bed of Scotland (based on data of Aldridge *et al.*, 1993); (b) restoration of soft parts surrounding the apparatus elements in the Early Devonian (Gedinnian) ozarkodinid conodont *Pandorinellina remscheiden-sis*; the head is shown in postero-ventral view and is "cut" to expose elements hidden in the throat; the anterior set of elements was oriented more obliquely toward the axis in resting position (from Dzik, 1991*b*); (c) proposed arrangement of elements and their homology in the Ordovician prioniodontid conodont *Promissum pulchrum* (the head is shown sectioned medially; Jeppsson's notation of element locations is modified by adding locations **ab** and **ho**, derived from "ambalodus" and "holodontus" element types of *Amorphognathus*, where they are not differentiated morphologically from elements in locations **oz** and **ne**); three-dimensional arrangement of elements taken from Aldridge *et al.* (1995). Note that in this position the apparatus could not effec-tively work, as some elements are located in between others. To act as a grasping apparatus, protrusion of the whole set of symmetry transition series elements ("filtra-tory basket") is necessary; this resulted in an arrangement of elements basically similar to that in the ozarkodinids.

1995) belongs to the Prioniodontida. Despite a rather large taxonomic and time difference between them, these conodonts are similar anatomically (Gabbott *et al.*, 1995). Both show lampreylike outline and well-developed ocular sclerotic rings and differ only in details of apparatus organization. However, these conodonts are highly derived and to restore the ancestral architecture of the conodont apparatus, which may be more meaningful than the rather uniform anatomy in the discussion of early chordate evolution, it is necessary to consider more primitive conodonts. Our understanding of such forms is primarily based on statistical reconstructions of their feeding apparatuses from collections of isolated elements and is subsequently refined by discovery of natural assemblages (for review see Sweet, 1988; Dzik, 1991*b*).

Restorations of the three-dimensional apparatus architecture of ozarkodinid conodonts represent the most reliable source of information regarding primitive conodonts. Findings of complete conodont apparatuses on the surface of bedding planes (natural assemblages) exhibit different directions of post mortem collapse, which enable restoration of the original spatial distribution of individual elements. Such analyses have been performed for the Early Devonian *Pandorinellina* (Fig. 1*b*; Dzik, 1976, 1986, 1991*b*) and several late Carboniferous genera (Aldridge *et al.*, 1987; Purnell, 1993; Purnell and Donoghue, 1997). Sometimes less deformed apparatuses have been recovered by acid dissolution of rock (clusters), where elements are fused together by diagenetic mineral. These provide additional evidence. An iterative method of reconstructing the original spatial arrangement of elements in the apparatus by proposing and then continually refining a model by testing with natural assemblages has proven to be extremely efficient. Little doubt remains that in the ozarkodinid apparatus elements were arranged in pairs, more or less transversely to the body axis, with their cusps opposed. Some minor discrepancies remain that refer to details of the arrangement of particular functional units within the apparatus (see Purnell and Donoghue, 1997). Only Nicoll (1995) continues to support a model where all elements are arranged parallel to the body axis, even though evidence occurs that is clearly contradictory (e.g., Aldridge *et al.*, 1987).

The number of element locations in all the ozarkodinid apparatuses appears to be equal. Also the morphology of elements in those locations are very similar. The late Paleozoic Spathognathodontidae, Polygnathidae, and Idiognathodontidae are a coherent morphological and biological group. Even further derived, Carboniferous species of the Prioniodinidae, representatives of which are well differentiated already in the early Silurian, do not seem to deviate from the ozarkodinid apparatus architecture. The restoration of a plan of their skeletal components proposed by

Purnell (1993) for *Kladognathus* slightly differs from that typical of more generalized members of the order, but it has not been based on assemblages preserving the original arrangement of elements but rather on functional assumptions derived from homology of elements.

Homology of particular element types in the ozarkodinid apparatus can be traced back in time to the late Tremadoc simple-cones *Rossodus* and *Utahconus* (see Dzik, 1991*b* for a review). The numerical ratio of element types seems to be consistent with the model of a 15-component element apparatus, although no statistical study of Ordovician collections has been performed since the early attempts by Marsal and Lindström (1972) and so is difficult to test.

The only direct evidence of prioniodontid apparatus composition (the second largest order of the conodonts, the Prioniodontida, most of which is early Paleozoic) is provided by the late Ordovician conodont *Promissum*. Natural assemblages show that this apparatus differed from those of the late Paleozoic, at least in the presence of two more element pairs: one of incisor-type and another of platform morphology (Aldridge *et al.*, 1995; Purnell *et al.*, 1995). The architecture of the apparatus of *Promissum* is otherwise consistent with that of *Pandorinellina* in the organization of the portion homologous to the symmetry transition series of more primitive conodonts, with one symmetrical and four paired elements (Fig. 1*c*). Numerical data on *Baltoniodus* from the Ordovician Mójcza Limestone of the Holy Cross Mountains (Dzik, 1994, 1996) suggests that duplication of the two element locations, which make prioniodontids different from ozarkodinids, had taken place by the Early Ordovician. After the Ordovician, only the pterospathodontids, which are probably derived from an ancestor close to *Complexodus* or *Icriodella* (Dzik, 1991*b*), flourishing in the early Silurian, had a similar number of elements to *Promissum* (see Männik and Aldridge, 1989).

It is possible to determine homology between apparatuses of ozarkodinids and the Protopanderodontida. The protopanderodontids range back to the latest Cambrian. They are almost certainly ancestral to both the Ozarkodinida and the Prioniodontida and probably, therefore, had a similar apparatus organization. This is supported by evidence from incomplete clusters (reviewed by Dzik, 1991*b*). These clusters do not allow determination of the exact number of element locations, but underrepresentation of the symmetrical elements in samples of isolated elements relative to other element types suggests that they were unpaired. The two robust element types in the apparatus probably correspond to elements of ozarkodinids and prioniodontids traditionally referred to as the platform series. Composition of the protopanderodontid apparatus was, therefore, similar to that of the ozarkodinids and the prioniodontids. However, this is not the primitive state of true conodonts.

Much lower diversification of coniform elements occurs in the apparatus of the order Panderodontida. The most complete cluster of the panderodontid elements, currently known, consists of 13 elements (Dzik and Drygant, 1986). The only unpaired element in this specimen is strongly asymmetric and was undoubtedly originally paired, the missing element probably lost during processing. All elements were, therefore, originally paired. In several genera of the panderodontid conodonts no symmetrical element was present in the apparatus. The cluster of *Besselodus* described by Aldridge (1982) consists of seven elements, representing one side of the apparatus in close lateral connection, an arrangement unlikely to have developed if the apparatus was similar in organization to that of ozarkodinids or prioniodontids. Even if symmetrical elements do occur in *Panderodus*, their occurrence in number comparable with other elements of the apparatus suggests that they too were paired. In some species they are so rare that they are more likely to have resulted from an atavistic malformation than to represent a normal element type. All of these facts are reconcilable with the chaetognathlike model of apparatus architecture proposed by Dzik and Drygant (1986). It has been proposed (Dzik, 1991b) that the primitive 15-element component apparatus can be derived directly from that of the Panderodontida by adding a single medial element to the existing seven pairs of elements. This suggests that the prioniodontid apparatus is derived. Sansom and colleagues (1995), who had at their disposal the only known natural assemblage of *Panderodus*, preferred to arbitrarily reconstruct an apparatus with 17 elements, more than present in the natural assemblage and in any known clusters. The maximum number of element types allowed by their model is nine, but the most observed is seven or fewer. Whether a 19-component apparatus, instead of 15, is primitive or derived is unknown, although at present the second possibility appears more likely.

Symmetrical elements occur in pairs in the apparatus of the Tremadoc conodont *Coeleocerodontus*, known from numerous clusters (Andres, 1988). Its apparatus consists of two pairs of relatively robust elements at one end, at least five pairs of the "symmetry transition series" (including large, paired symmetrical elements), and a pair of elements with a triangular cross-section at the other end. The exact affinity of *Coeleocerodontus* remains obscure (Andres, 1988 suggested that it was a paraconodont, a member of the Westergaardodinida, an order defined on element histology), but its apparatus structure is similar to that of the protopanderodontids. However, occurrence of all elements in pairs suggests that

Coelocerodontus is more closely related to the panderodontids. Perhaps the apparatus of *Coelocerodontus* is representative of the condition from which both the simple panderodontids and more advanced protopanderodontids are derived.

FUNCTION OF CONODONT APPARATUSES

The strong morphological differentiation of the conodont feeding apparatus mirrors the differentiation of element function. The clear analogy between the organization of the conodont apparatus and the dentition of mammals (with posteriorly located molars and anterior canini), as well as the crustacean mandibles (with the anterior incisor, posterior molar regions, and comblike lacinia mobilis in between), strongly suggests a feeding function. Crustacean mandibles are closest in size and in morphology to conodont apparatuses and provide a good functional analog. The presence of the unpaired element in the middle of advanced conodont apparatuses makes their anterior and posterior parts functionally distinct. The anterior set of sharply denticulated elements was probably exposed and could have been used to grasp and possibly perform a filter function, whereas the more robust posterior elements were hidden in the throat of the animal, performing grinding functions.

Although in general this functional interpretation of the conodont apparatus seems well substantiated, it is much more difficult to be more precise about the functional adaptations of particular element types. Purnell (1994) used the positive allometry in development of the platform area in the posteriormost elements in the apparatus of the idiognathodontids and the allegedly negative allometry in the relative growth of the anteriormost elements (the latter being clearly an artifact resulting from the chord instead of the true length of the curved inner process being measured) to support their crushing and grasping functions, respectively, for these elements. Although such a functional differentiation of these elements is generally accepted (Jeppsson, 1979), this does not necessarily imply their morphology being so strictly controlled histogenetically. The histogeny of conodont elements to some degree is recapitulated in their evolution. The reason for the difference between juveniles and adults is historical rather than functional. Both the platform and the icrion (the two basic ways of molarization in conodonts—see Dzik, 1991*a*) are evolutionarily derived structures that in all conodonts originated late in histogeny, gradually expanding in the course of evolution toward earlier and earlier ontogenetic stages. In advanced conodonts (e.g., *Manticolepis* or similar palmatolepidids) growth of the platform is isometric except for the earliest phase of histogeny.

The mammalian jaw or crustacean mandible models are not applicable to the most primitive panderodontid apparatuses, in which the chaetognath grasping apparatus appears to be a much closer analog (Dzik and Drygant, 1986). It has to be noted, however, that the most primitive coniform elements of the Late Cambrian conodonts were generally of more robust appearance than chaetognath grasping spines (including the associated protoconodonts, which probably belonged to the chaetognaths; Szaniawski, 1982). The apparatuses they formed were thus probably relatively robust as well.

GROWTH AND FUNCTION OF CONODONT ELEMENTS

The unique internal structure of the elements of the conodont apparatus is a result of their mode of growth, with successive lamellae of the crown being added from outside. Frequently, within the conical basal cavity (or just below the crown if the basal cavity is inverted) an additional, histologically different kind of skeletal tissue was secreted. The nature of this basal filling tissue, being secreted from inside the basal cavity, was established by Gross (1957). Lindström and Ziegler (1971) have shown that the basal cavity tissue in evolutionarily primitive conodonts was originally rich in organic matter, or was plainly organic in composition, and its mineralization was commonly secondary. Generally, gerontic elements have their basal filling tissue better developed than juvenile ones. Elements of the same species may or may not bear any basal body, depending on the sample. Samples rich in phosphatized fossils frequently yield elements with elaborate structures of this kind, which supports the notion of a significant contribution from a secondary mineralization, but there are many exceptions and the variation in mineralization of the basal bodies remains to be explained.

The lower surface of the basal filling tissue in early Ordovician conodonts is of irregular appearance, with perpendicular branching canals opening to the basal cavity (Lindström and Ziegler, 1971: pl. 4: 6) interpreted as at least analogous to dentine tubuli (Dzik, 1986: Fig. 1a; Sansom et al., 1994). Sometimes the tissue is spongy, black, and probably unmineralized (Dzik, 1986: Fig. 1b). These features of the basal filling tissue are even more apparent in Cambrian conodonts (Andres, 1988; Fig. 2b, this chapter).

The only developmental analog of the crown and basal filling tissues of conodonts (with respect to their mineralogy and mode of secretion) are

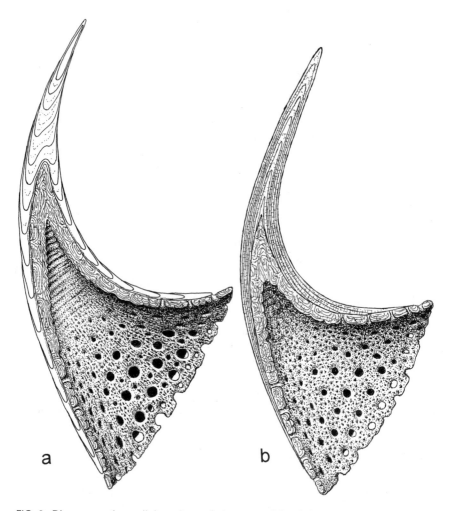

FIG. 2. Diagrammatic medial sections of elements of Cambrian "paraconodonts" and "euconodonts" (based on Andres, 1988): (a) the Late Cambrian westergaardodinid *Problematoconites* sp. (see Andres, 1988: text—Fig. 13, pl. 4: 1–8), note initial rodlike stage in the development of the crown tissue; (b) a primitive late Cambrian cordylodontid (see Andres, 1988: pl., 12: 7–8, 13: 5–6), the crown is conical from its beginning.

the enamel and dentine of vertebrates (Schmidt and Müller, 1964; Dzik, 1976, 1986; Smith, 1995; Sansom, 1996). Since the chordate nature of the conodonts has been established, there have also been other attempts to fit the conodont element microstructure into the classification of vertebrate skeletal hard tissues. Sansom *et al.* (1992) proposed that the "lacunate spaces" within opaque parts of conodont crown tissue ("white matter") were osteo-cyte lacunae and thus the tissue was a cellular bone. Secondary develop-ment of white matter during late stages of the element's histogenesis (or perhaps even diagenetically), replacing the original lamellar cone-in-cone structure, is well recognized (Sweet, 1988, p. 14; Donoghue, 1998, Fig. 6, pro-posed that the white matter was developed at the cusp surface during early histogenesis but this is in contradiction with the observed pattern of its dis-tribution in juvenile elements). There cannot, therefore, have been cells within the tissue during secretion. Furthermore, as white matter is invari-ably covered by intact lamellar crown tissue, there is no way for the puta-tive cells to migrate there (Dzik, 1993). Moreover, the external secretion of the crown tissue makes it rather unlikely to be of ectomesenchymal or mesodermal origin, which is the case with the cellular bone.

The claim that the basal filling tissue of the Ordovician *Chirognathus* is dentine (Sansom *et al.*, 1994) is consistent with the homology proposed by Schmidt and Müller (1964), but usually canals comparable with dentine tubuli are missing in the basal filling. There is also no clear evidence yet that mineralization of the ectomesenchymal organic tissue filling the basal cavity characterized the common ancestor of conodonts and agnathans. This would be necessary to identify the basal filling tissue as dentine. They thus have to remain just analogs.

Ganoid scales of Recent fishes (see e.g., Meunier, 1980) provide prob-ably the closest analog, if not homolog, of the conodont element mineral tissues. The ganoine is a primitive type of enamel (Sire, 1995). The cells of the inner ganoine epithelium are homologous with ameloblasts and can be referred to as such (Sire, 1994). This can be reasonably extended to the secretory cells of the conodont crown tissue. Ganoine secretion proceeds by alternating deposition of a thin organic layer of preganoine, and a sub-sequent period of mineralization. As a result, distinct growth lines develop at the contact of ganoine with the underlying bony plate. Instead of colla-gen, preganoine contains fibers of other proteins arranged perpendicularly to the surface, along which ganoine crystals grow (Sire, 1995). It is likely that the histogenesis of conodont elements (e.g., Müller and Nogami, 1971) was homologous, and growth was punctuated by periods of arrest (perhaps daily; Fig. 3d; see Zhang *et al.*, 1997). Kemp and Nicoll (1995) argued for the presence of collagen in the conodont element tissues because they stain with Sirius Red, the tetrakisazo dye used in soft tissue histology. It seems to be of importance in this context that simple histological tests failed to identify even structurally preserved Early Paleozoic collagen (Urbanek, 1986) whereas in phosphatic sclerites of Ordovician palaeoscolecid worms, mineralogically similar to conodont elements, collagen fibers are represented only by empty spaces (Dzik, 1986). Sirius Red acts by ion

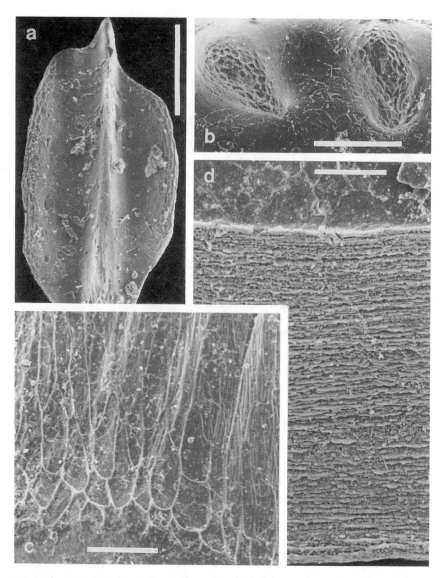

FIG. 3. Localized development of secretory cell imprints (a, b), their elongation with denticle growth (c), and growth increments (d) in conodont elements: (a) juvenile **sp** element of *Polygnathus* s. l. from the Late Devonian (latest Frasnian) of Płucki, Holy Cross Mountains, Poland, in occlusal view showing development of polygonal cell imprints at margins of fast growing platform; bar scale 49 μm; (b) juvenile **sp** element of *Icriodus alternatus* from coeval strata in Wietrznia, same region, with polygons developing on fast-growing denticle tips of icrion; bar scale 49 μm; (c) adult **hi** element of *Falcodus* sp. from the Early Carboniferous (Tournaisian, sample Dz–15) of Dzikowiec, the Sudetes, Poland, in lateral view showing elongation of secretory cell shapes toward the denticle tips; bar scale 19.6 μm; (d) surface of inverted basal cavity in **ne** element of *Ancyrodella* from the Late Devonian (late Frasnian, sample Pl–22) of Płucki showing numerous rhythmic, possibly daily, increments; bar scale 19.6 μm.

binding of its sulfonic groups with the basic groups of the collagen molecule (Marotta and Martino, 1985), mainly proline and hydroxyproline, and is also known to stain several other kinds of proteins (Brigger and Muckle, 1975).

Although some irregularities have been documented from the internal record of growth of conodont elements (Müller and Nogami, 1971), well-preserved conodont apparatus elements, like scales of pelagic fish, exhibit no evidence of surface wear (scratches and abrasions, such as illustrated by Purnell, 1995b, are uncommon). An active grasping and crushing function of the conodont elements has been inferred from their shapes (Jeppsson, 1979). If the mineral crown tissue of the elements were in a direct contact with food particles this would hardly allow preservation of such intact surfaces unless they were periodically covered by folds of soft secretory tissue, as mentioned by Denham (1944) and as fully elaborated by Bengtson (1976). The pocket theory of secretion fits some peculiarities of the morphology of certain conodont elements (Carls, 1977) but is difficult to apply to complex brush-like morphologies of the occlusal surfaces of some elements and difficult to test. However, some conodont elements record the arrangement of secretory cells on their oral surface, which provides direct insight into the anatomy of the organ responsible for their secretion.

Such imprints were first recognized by Hass (1941), and their value for paleobiological analysis is widely recognized (Pierce and Langenheim, 1970; Burnett, 1988; Conway Morris and Harper, 1988; Burnett and Hall, 1992; von Bitter and Norby, 1994; Zhuravlev, 1994). Their width, usually ranging from 3 to 5 µm, is similar to that of the cells of inner ganoine epithelium (Sire, 1994) and is even closer to ameloblasts in Recent tetrapods where they are of columnar shape, with a height of about 50 µm, and with well-developed Golgi apparatuses (Masuda et al., 1989) crucial to their secretory activity (Matsuo et al., 1992). In conodonts, calcium phosphate was secreted by the whole exposed surface of the cells, which were individually convex at the secreting margin. Where cells were in contact with their neighbors, activities of the cells overlapped, leading to increased deposition of calcium phosphate. This resulted in the development of a polygonal pattern of ridges on the surface of the crown tissue. Not only the size, but also shapes of individual cells can be traced through histogeny (Burnett, 1988) and it appears (Fig. 4) that they were not uniform over the whole area, as would be predicted if the elements were temporarily withdrawn from epithelial pockets (or periodically worn-out and regenerated). Burnett (1988, pp. 414–415) salvaged part of the theory, proposing that "more positively upstanding areas" were emergent during function. His hypothesis is refuted by the presence of polygons on the tips of cusps in icrions of genera

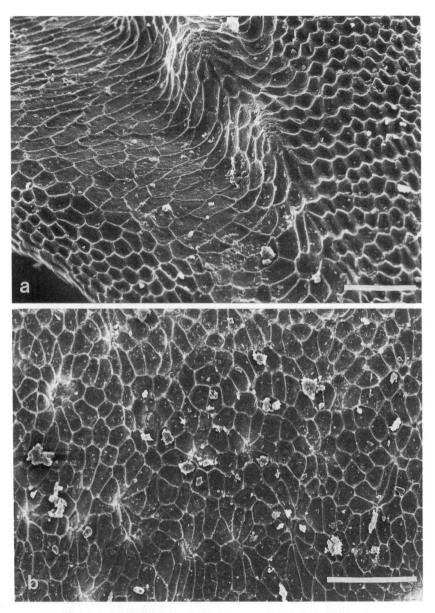

FIG. 4. Distribution of ameloblast imprints on the occlusal surface of the posterior-most (**sp**) elements of the conodont apparatus: (a) *Siphonodella belkai* from the early Tournaisian of Kowala, Holy Cross Mountains, sample Ko–42; note that the epithelial cover passes uninterrupted across the element carina (the main row of denticles) and that their shapes vary specifically; bar scale 19.6 μm; (b) *Conditolepis? linguiformis* from the latest Frasnian of Płucki, Holy Cross Mountains; note star-like arrangement of cells at tubercles; bar scale 49 μm.

such as *Icriodus* (Fig. 3*b*) or *Lochriea* (von Bitter and Norby, 1994) and on the lateral surfaces of denticles in, among others, *Dinodus*. Cells on the lateral surfaces of denticles were extremely elongated (Fig. 3*c*; contra Burnett, 1988, p. 414). The most reasonable explanation for this pattern is the occurrence of extensional stress generated by the elevated denticle tip during growth, whereas the cell divisions on the lateral surfaces of the denticle were, for some reason, inhibited.

The mode of secretion of the crown tissue in conodonts that develop a prominent pattern of reticulation, sometimes with deep concavities in the center of each polygon (Fig. 4*a*), closely resembles the mode of secretion of enamel in Recent tetrapods (e.g., Boyde, 1978). Conodont crown tissues (for instance *Polygnathus* s. s.), which did not develop polygons, even in the thickest parts of the platform, more closely resemble the ganoine of Recent primitive actinopterygian fishes (e.g., Sire, 1994, 1995). It is possible that in such cases, secretory cells were not in direct contact with the mineralized tissue but were separated from it by a membrane similar to the ganoine membrane. This is a kind of basement membrane, which does not mineralize, and it is rich in proteoglycans that work as a glue keeping epithelial cells in place (Sire, 1995). Presumably in places of extremely high secretion of conodont crown tissue (Fig. 3*a*, 3*b*), the membrane tended to thin or disappear completely. As a result, direct contact of the epithelial cells (ameloblasts) with the secreted tissue led to the formation of a polygonal surface pattern.

In platform elements, concave areas, which had a relatively low secretion rate, are virtually always smooth, lacking imprints of secretory cells. The boundary between areas of smooth and polygonal ornament are distinct, or can be seen as gradual over a very short distance, where the ribs delimiting polygons become more and more indistinct, as if the surface is covered by a thin blanket. This strongly suggests the presence of a basal membrane separating the mineralized matrix and the secretory cells in the areas of the smooth surface. Microstriae also present on the surface of such elements have nothing to do with cell imprints but represent longitudinal folding of the basal membrane, as shown by irregularities in their distribution. The occurrence of cell imprints discernible even in areas such as these indicates that the cells were invariably covered by a secretory epithelium, elements differing only in the extent of the basal membrane. The membrane, therefore, tends to disappear in the fastest growing parts of elements. There is an interesting analogy between this pattern of the secretive epithelium growth and the morphogenesis of salivary glands, as reviewed by Bard (1990, pp. 44–48). After a bulblike bud of the salivary gland covered by a basal lamina is formed, the surrounding mesenchyme produces a neutral hyaluronidase that locally degrades the chondroitin sulfate (one of several sulfated proteoglycans of an important function in morphogenesis) and the hyaluronic acid of the lamina. Only regions where collagen is attached to the basal lamina are protected and they eventually become the concave areas surrounding the faster growing protrusions.

Reticulate imprints of the secretory cells that developed at the ends of discrete periods of growth are present in cross-sections of the Carboniferous *Siphonodella* elements (Burnett, 1988). This provides additional support to the idea that periods of growth of the conodont element corresponded to sequential secretion of an organic matrix and its subsequent mineralization.

The most important feature of ameloblast distribution in conodonts is that they provided continuous and uniform cover over the whole of the occlusal surface (Figs. 4a, 5) making Bengtson's (1976, 1983) pocket theory untenable. This leaves Schmidt's (in Schmidt and Müller, 1964, p. 131; also Priddle, 1974) "horny cup" model the viable paradigm to explain the paradox of growth and function in conodont elements. The epithelial cover of the elements had to perform mechanical functions of the feeding apparatus (Fig. 6). The covering of epithelium over the surface of conodont elements must have been multilayered, as in chaetognaths and in virtually all of the chordates, and may have been produced in a similar manner to keratinous toothlets of *Myxine* and *Petromyzon*. Chordates, like other deuterostomes, lack chitin synthetase (P. Willmer, 1990, p. 80), so whenever a need to develop a firm external skeleton appears, its function is performed by keratinized epidermal cells.

Functional convergence between conodont and cyclostome feeding apparatuses has been invoked by many authors (e.g., Priddle, 1974; Dzik, 1986; Sweet, 1988; Krejsa et al., 1990a), but if a homology can be drawn with the toothlets of myxinoids, it must be restricted to the unpreserved (hypothetical) horny caps of conodont elements and not the crown tissue of conodont elements (contra Krejsa et al., 1990a, b; also Slavkin and Diekwisch, 1996). The structure of the horny caps remains enigmatic, although Schmidt (in Schmidt and Müller, 1964, p. 130) refers to the presence of black stains extending from the tips of denticles in natural assemblages of *Gnathodus* from the Early Carboniferous of Germany. Whether these are the remnants of keratinized epidermal cells (as in *Myxine*) or a chitinous laminar structure secreted by epidermal cells (as in chaetognaths) cannot as yet be determined, although the second possibility seems unlikely because chordates are not able to secrete chitin. Obviously, phylogenetic implications of this distinction would be far reaching, but either structure could have evolved from the other by a change in the secretional behavior of the epidermal cells, from an ability to secrete an external proteoglycan–polysaccharide matrix to a modified keratinous cytoskeleton, or vice versa.

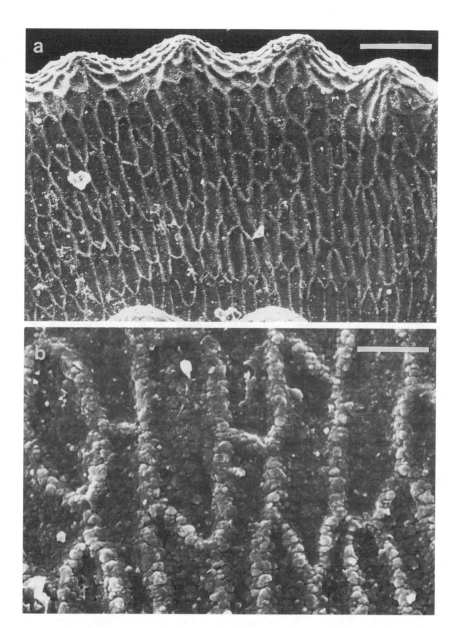

FIG. 5. Relationship between calcium phosphate secretion rate and the size of ameloblasts in *Siphonodella belkai* from the early Tournaisian of Kowala, Holy Cross Mountains, sample Ko–42; (a) platform surface in the specimen illustrated on Fig. 4a. The platform margin with tubercles was secreted at the highest rate and ameloblast contact areas are the smallest. At the platform occlusal surface the secretion was much less intense, ameloblast contacts are much wider, and are also mechanically extended in accord with the platform widening; scale bar 19.6 μm; (b) a few ameloblast contact areas at higher magnification showing arrangement of calcium phosphate crystals; scale bar 4.9 μm.

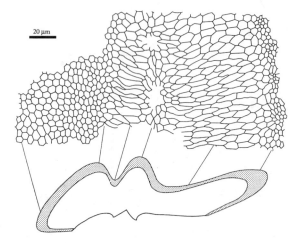

FIG. 6. Ameloblast shapes in the specimen of *Siphonodella belkai* illustrated on Fig. 4*a*; contours of ameloblast imprints are traced from SEM photographs taken at different angles, to keep views of particular cells vertical. The intensity of secretion corresponds to the thickness of laminae in the crown tissue. It would be technically difficult to section exactly the same element, at the surface of which the cell imprints have been traced (and rather unreasonable to destroy it in result) so only a rough estimate of secretion rate can be inferred from the element's shape and its isometric (in short-term) growth.

To avoid a contradiction with the data on the secretory cells imprints discussed previously, it seems reasonable to consider the wear, common on denticle and tubercle tips in isolated conodont elements (Purnell, 1995*b*), as a post mortem taphonomic abrasion.

CONODONT ELEMENT MORPHOGENESIS

The elements of the conodont oral apparatus attain their specific shapes as a result of differential rates of secretion of calcium phosphate by the ameloblasts. Presumably, the ameloblasts first secreted a thin layer of organic matrix followed by precipitation of calcium and phosphate ions, forming crystallites that developed along the axes of protein fibers in the matrix, as in the formation of enameloid (Shellis, 1978; Prostak *et al.*, 1993) and ganoine (Sire, 1994, 1995). The protein fibers are removed from the matrix while it is mineralized. In conodonts, secretion was the most intense (at least during early stages of the histogeny) at the cusp. Most of the elements also bear ridges or crests that arm the inner and outer margins of the cusp, which frequently bifurcate or are supplemented by additional ridges and crests. At the element margins, the ridges tended to develop element rami (processes) that often dominated the whole element. Typically, after the processes reach a certain size, the sharp occlusal margins develop denticulation. The basal margins of conodont elements are approximately flat, the occlusal morphology resulting exclusively from differential rates of crown tissue secretion.

The ameloblast imprints also provide an insight into the mode of secretion of specific structural features of conodont elements. They show clearly that the size of the contact area of ameloblasts with the element crown tissue surface is inversely proportional to the intensity of secretion. The polygons are smallest at the tips of denticles and on the elevated margins of platforms (Figs. 5, 7). In such areas, the cell imprints are deep and, in extreme cases, a small depression may occur in the center (Fig. 4a) that may correspond to Tomes' process in mammalian enamel. This feature is

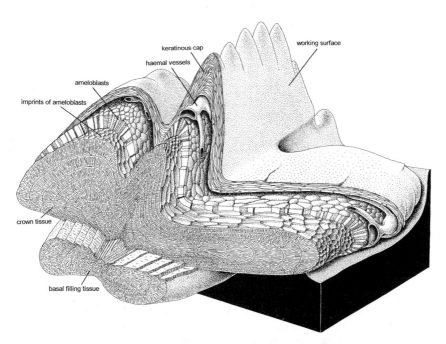

FIG. 7. Hypothetical section through the secreting organ of the conodont element, exemplified by the **sp** element of Tournaisian *Siphonodella*. Shapes of ameloblasts have been inferred from their imprints at the element surface (Fig. 4a), the presence of a blood lacuna above the cusp from the inferred control of secretion, the horny cup is required to provide enough strength for biting. Basal filling tissue in conodonts of this group is generally unmineralized but perhaps some structureless organic connective tissue was present in the corresponding place.

present on the tips of rounded tubercles, for instance in *Lochriea* or in *Sweetognathus*.

In elements with wide platforms that develop surface denticulation (e.g., in palmatolepidids), each minor tubercle is marked by an asterisk-shaped radial cell arrangement, commonly with a single minute cell imprint or group of imprints in the center. This suggests that the development of such denticles was preceded by a proliferation of ameloblasts. In higher vertebrates the proliferation of ameloblasts is stimulated by endogenously produced epidermal growth factor (EGF; a low molecular weight polypeptide; see Soler and Carpenter, 1994) that stimulates DNA synthesis (Hu *et al.*, 1992). Tooth cusp formation commences with the development of a cluster of nondividing epithelial cells (enamel knot), which stimulates the growth of surrounding cells by synthesis of fibroblast growth factor FGF–4 (Jernvall *et al.*, 1994), one of the family of single chain polypeptides with evolutionarily conservative structure (see Vainikka *et al.*, 1994). Both of these growth factors transduce their signals by binding to cell surface receptors exhibiting tyrosine kinase activity. The cells of the enamel knot undergo apoptosis after ceasing their regulatory action (Thesleff and Nieminen, 1996). Perhaps the minute polygons at the tip of the tubercles and denticles in conodonts correspond to cells located immediately below an enamel knot. It has to be noted, however, that during growth of the conodont crown there is no apparent increase in the number of secretory cells (von Bitter and Norby, 1994), which were inserted to accommodate the increase in area resulting from growth. One may speculate that a high secretion activity requires DNA replication processes to have ceased. Shape distribution remains constant during growth although adult elements of *Siphonodella* exhibit thicker ribs separating polygons that tend to constrict the contact area, particularly of laterally elongated cells in areas of reduced secretion rate.

A small area of surface contact does not necessarily mean that the cell itself was small. The cells putatively located immediately below the enamel knot could have been tightly arranged, vertically elongate, and columnar, like mammalian ameloblasts. This has already been suggested in connection with the inverse relationship between cell size and secretion intensity. Cells that were laterally elongate, and presumably low in profile, did not contribute much to element growth. Perhaps the cells were equally capable of calcium phosphate secretion and reduced contact area may, therefore, have enabled more intense secretion. The tip of each conodont ameloblast appears to be a functional analog of the Tomes' processes of mammalian ameloblasts, capable of secreting crystallites perpendicularly to its surface (see Masuda *et al.*, 1989).

If this was truly the case, tubercle and denticle morphogenesis was almost exclusively controlled by the supply of a morphogenetic factor, phosphate, and calcium. Retinol (vitamin A) may work in tooth morphogenesis as a pattern-generating factor that acts by increasing expression of EGF mRNA (see Smith and Hall, 1993, for review); transferrin is also recognized as the serum protein that is a necessary growth factor for early tooth morphogenesis in vertebrates (Partanen and Thesleff, 1989). A localized increase in the rate of secretion of calcium phosphate requires that the ameloblasts are supplied by other organs of the body. The transport direction was obviously from the element base to its tip, whatever model of the element anatomical organization is assumed. If this process proceeded exclusively by active transport through cell membranes (as is probably the case in the ganoine of *Polypterus*, which is secreted very slowly; see Sire, 1995), the normal gradient would be opposite to the observed secretion intensity. During a presumably short period (not more than a few years; Zhang *et al.*, 1997) each of the cells had to secrete an amount of calcium phosphate several times larger than its volume! It therefore appears reasonable to suggest that some blood sinuses developed above the ameloblasts at tips of denticles to supply the most active ameloblast, and that theses canals were responsible for the supply of food and the stimulation of mitosis and cell fission. Such a network develops above the ganoine epithelium in the Recent fish *Lepisosteus* (Sire, 1994). The development of a sinus preceded, according to this model, the origination of denticles. How tooth (and probably also conodont element) organogenesis is precisely executed remains largely unknown, but the homeobox gene Hox–8 (for a review of homeobox genes in primitive chordates see Holland *et al.*, 1994; Holland and Garcia-Fernandez, 1996) is definitely involved in specifying tooth shape and is expressed in the enamel knot and the epithelium (MacKenzie *et al.*, 1992).

The inferred mode of the morphogenetic control of element shapes suggests the presence of a blood sinus (or an expansion of a capillary) above each area of elevated morphology on the surface of an element. This means that each of the denticles in an element had to have been capped by a soft, fluid-filled cavity, which makes the idea of protective horny caps even more appealing. In fact, the tip of the epithelial papilla, at which the sharp myxinoid teeth (built of keratinized cells; Slavkin and Diekwisch, 1996) or the chaetognath grasping spine develops (composed of α-chitin; Bone *et al.*, 1983; Kapp, 1991), must also be a center of increased skeletal matrix secretion. A way to stimulate the supply of nutrients to exactly the same place has thus to be assumed also for the horny cap.

If this interpretation is correct, the morphology of conodont elements relates directly to morphogenesis of the blood vascular system or otherwise had a reciprocal relationship with it. This requires that dermis was inserted between the external epithelium and the ameloblastic layer, as is the case in vertebrate teeth and in ganoid scales. The inferred presence of a horny cap above the secretory cells requires higher demands on the oxygen supply than at the surface of fish scales where epithelium is directly exposed to the sea water. It would be expected that in such conditions the distance between blood capillaries was less than 100 μm, as is typical of Recent vertebrates, which corresponds quite closely with the distances between ridges in conodont elements. This is exemplified by the histogeny of elictognathid conodonts (Dzik, 1997), where new ridges were developed with uniform spacing and terminate at a level of the cusp (even if the cusp itself is completely reduced). A similar morphogenetic field controlled by an angiogenetic mechanism was probably executed at early stages of the development of the apparatus, after buds of particular elements had been established. The way in which endothelial cells form new blood capillaries possibly explains the bifurcation of processes and ridges in conodont elements (see Dzik, 1994).

Although, in some cases, the distribution of denticles on the surface of conodont element platforms seems to be almost random, this was not the case with the pattern of early denticle growth. A well-defined morphogenetic field was probably generated by a denticle formation centrum (blood lacuna?) that inhibited formation of new denticles in close proximity. The field size increased during the histogeny. This might explain some peculiarities of conodont element growth where earlier formed denticles coalesced and new individual denticles were periodically added to the growing end of an element (Fig. 8; see Dzik and Trammer, 1980).

One may speculate that new denticles started to develop only after the element process base reached a sufficient length to be free of the inhibitory effect of the morphogenetic field of preceding denticles. Only then could a new blood sinus have developed, an ameloblast below which must have begun to proliferate and develop a group of densely packed cells with small secreting areas. Local increase in the secretion of calcium phosphate would then have resulted in elevating the tip ameloblasts and in the mechanical elongation of their neighbors. As a result a rosette pattern of ameloblast distribution would have developed (Fig. 4b), and it would have subsequently been obliterated when a denticle developed with smooth sides and a sharply pointed tip.

Therefore, it appears that although conodont hard tissues can easily be homologized with the hard tissues of vertebrate dermal denticles, conodont elements were functionally rather unlike such denticles and teeth. The only possible homolog of the inferred mechanically functional organic cap is the horny teeth of the hagfish (Priddle, 1974). In fact, the genes responsible for amelogenesis are universally distributed among the vertebrates, including the myxinoids (Slavkin and Diekwisch, 1996), and it is reasonable

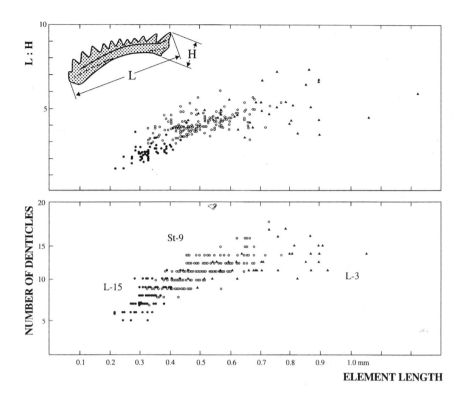

FIG. 8. Scattergrams showing the pattern of conodont element growth as exemplified by **sp** elements of the Middle Triassic *Gondolella mombergensis-G. haslachensis* lineage (from Dzik and Trammer, 1980); specimens are from samples Lesica 3 (*G. mombergensis*; the oldest one), Stare Chęciny 9 (transitional), and Lesica 15 (*G. haslachensis*); the evolution leads toward more and more juvenile morphologies dominating in samples: (a) elongation index (element length: cusp height) against the element length; note that until elements reached a certain size their growth was mostly due to an increase in elongation; (b) number of denticles plotted against element size; element growth proceeded through addition of more and more denticles at the ventral ("posterior" in conventional terminology) end until coalescing of earlier produced small denticles counterbalanced this.

to assume that their ancestors had abilities to secrete enamel. Perhaps these were *Archeognathus*-like conodonts (Dzik, 1986; see also Klapper and Bergström, 1984). The Late Carboniferous *Gilpichthys* shows some features that fit this interpretation in its dorsally attached mouth apparatus of keratinous teeth, six gill pouches, and V-shaped myomeres (Bardack and Richardson, 1977). All Recent vertebrates show rather W-shaped myomeres (Pridmore *et al.*, 1996), but it remains unclear whether this character, generally connected with increased locomotory abilities, developed only once in the chordate phylogeny. Another fossil chordate that may be related to this lineage, although it had only a pair of conical teeth, is the Early Carboniferous *Conopiscius* (Briggs and Clarkson, 1987). Recent hagfish have developed a powerful muscular system to operate their jaws (Dawson, 1963; Yalden, 1985) that corresponds to the rather robust appearance of their teeth. Such a muscular armament was apparently missing in the typically gracile oral apparatuses of conodonts and its development must have been correlated with an increase in the thickness of their horny caps that perhaps made the internal phosphatic skeleton unnecessary.

Conodonts inhabited various marine environments, mostly pelagic but also reefal or extremely shallow water, but in their long-lasting evolution they never invaded brackish or fresh waters. This indicates that they were not equipped with osmotic regulation organs (renal tubuli) and their blood was apparently isotonic to the sea water. This places them physiologically at the same level as hagfishes, in which tubuli are virtually missing (Fels *et al.*, 1993), but below lampreys and other vertebrates.

In any case, the conodonts were true chordates. Their most obvious chordate character is the V-shaped arrangement of their myomeres (Briggs *et al.*, 1983). Such an arrangement also characterizes the relatively poorly known *Pikaia* (Conway Morris and Whittington, 1979; Whittington, 1985) and *Metaspriggina* (Simonetta and Insom, 1993) from the Middle Cambrian Burgess Shale. Little is known about the anatomy of the oral area of these animals but at least *Pikaia* shows some tentaclelike oral appendages (Conway Morris and Whittington, 1979; Whittington, 1985), which in *Nectocaris*, a possible close relative, may have been scleritized (Conway Morris, 1976*b*). The crucial role of the neural crest in the development of teeth in Recent vertebrates is used as evidence of their origin in chemosensory organs (Smith and Hall, 1990; see also Mallatt, 1996). Perhaps the oral structures in these Cambrian chordates represent such organs, which also performed a grasping function and may be homologous to the oral apparatus of conodonts. An oral apparatus was also present in *Yunnanozoon*, possibly the most primitive chordate.

RELATIONSHIPS OF *YUNNANOZOON*

Yunnanozoon is distinct from both Recent and Cambrian chordates and is presumably more primitive in its anatomical organization. It differs mainly in the dorsal disposition of its metameric blocks, or chambers, of which there were only 23 (Hou *et al.*, 1991), separated by straight vertical myocommata (Figs. 9, 10). They are thus of much simpler organization than

FIG. 9. *Yunnanozon lividum*; Early Cambrian Chengjiang fauna, Yunnan, China, juvenile specimen ELRC 52002, level 5; note the notochord, well delimited with dark lines, located clearly above the sediment-filled pharynx and intestine (contrary to Shu, Zhang, and Chen, 1996; Shu, Conway Morris, and Zhang, 1996), and the laterally compressed mouth apparatus of sclerites arranged in a ring (here laterally compressed). Scale bar equals 4.5 mm (from Dzik, 1995).

in Recent amphioxus. The complex topology of myotomes in higher vertebrates clearly corresponds to their locomotory abilities; there is an evolutionary tendency toward more sophisticated geometries (e.g., Hardisty and Rovainen, 1982, Fig. 11), and the most ancient chordates should be the simplest in this respect. Almost all of the known specimens of *Yunnanozoon* (more than 60) are preserved lying on one side–strongly suggesting that its

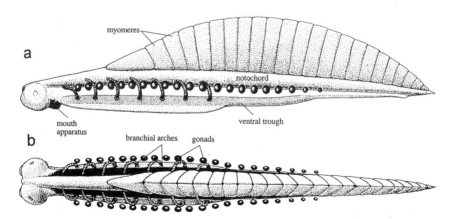

FIG. 10. *Yunnanozon lividum*; Early Cambrian Chengjiang fauna, Yunnan, China, restoration of preserved body organs (from Dzik, 1995).

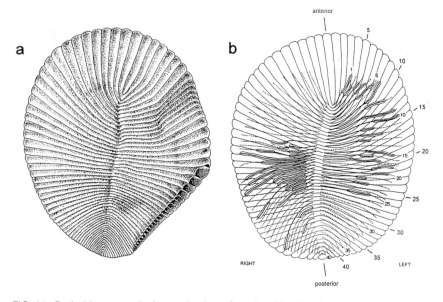

FIG. 11. Probable anatomical organization of the late Vendian *Dickinsonia costata*; (a) arrangement of dorsal chambers based on the specimen from the Ust'Pinega Formation of Russia (based on the specimen illustrated by Fedonkin, 1987); (b) diagrammatic representation of sand-filled metameric caeca of the intestine and dorsal muscular chambers as preserved in the specimen of Ediacara illustrated by Long (1995, p. 15). Boundaries of the dorsal muscular chambers and contours of sand-filled caeca are given in thin lines, the course of caeca inferred from their sand fillings shown as thick lines, their hypothetical occurrences being dotted. Caeca and muscular chambers are consecutively numbered on the left side of the specimen. The caeca diverge at the proximity of the inferred body wall (segment 7 and 8 on left side;) or merge close together (segment 5–7 on left), which means that they were not bound by any mesenteria.

body was laterally compressed. Immediately ventral to the muscular chambers there was a very large notochord that extended to the anterior end of the body (Fig. 9; Chen *et al.*, 1995; Shu *et al.*, 1996a claimed that the "supposed notochord appears to have gut contents," which is not the case in any of the studied specimens with this structure, clearly delimited by dark lines, preserved; see Fig. 9). Its weak staining in fossils is suggestive of its being constructed from highly vacuolarized tissue. However, whether the stiffness of the notochord was controlled by the muscular action of stacked cells (as in *Amphioxus*; see Flood, 1975; for review see Jefferies, 1986; Gee, 1996; and Ruppert, 1997) or whether the cells within the sheath were arranged irregularly and, therefore, nonmuscular (as in larval vertebrates) remains unknown. The pharynx in the anterior part of the body was rather narrow and was bordered on both sides by seven branchial arches, each composed of minute segments, about 20 in number. The arches were convex in lateral aspect and somewhat inclined anteriorly in the dorsal region. The dorsal end of each arch was attached to the midlateral surface of the notochord and was ventrally connected to the ventral trough. The apparent external position of the arches in *Yunnanozoon* suggests that its gill rays were oriented toward the pharynx. This requires that the branchial apparatus of *Yunnanozoon* was dissimilar in organization to the branchial basket of the cephalochordates and the tunicates. Each segment of the arches may have supported its own gill blade. A transverse striation of similar density (approximately 0.15 mm) occurs on longitudinally arranged ridges in the only, incompletely preserved, nematode-like specimen, on the basis of which Shu *et al.* (1996*a*) proposed the new Chengjiang chordate *Cathaymyrus*. Its segmented body exhibits two parallel lines in the middle that means, irrespective of their interpretation, that either the animal was of a round cross-section or it is dorsoventrally compressed. Anyway, it seems premature to assume its chordate affinities or to draw any definite conclusion on the basis of such a problematic specimen—more convincing material is needed.

The pharynx of *Yunnanozoon* was large, probably expanding laterally into gill pouches between arches. The pharynx continued into a straight gut containing helically coiled fecal content. This may indicate the presence of a spiral valve (Shu *et al.*, 1996*b*), which commonly occurs in lower vertebrates. The anus was located in a position similar to that in *Branchiostoma*.

The paired oval bodies, regularly distributed along the body (perhaps corresponding in number to myomeres), resemble the gonads of *Branchiostoma* in their oval shape, high content of organic matter, and metameric distribution restricted to the proposed pharynx and were located outside the gills, immediately below the skin (Dzik, 1995). This closely resembles the organization of the branchiogenital region in the enteropneusts, and like in the enteropneusts, the gonads probably opened to the outside, instead of into the coelom, as in Recent vertebrates, where the gonads are separated from outside by a thick wall of body musculature.

Yunnanozoon also bore a ventral skeletal structure that supplemented the notochord (Dzik, 1995). It had a troughlike shape with angular ventrolateral margins, possibly equipped with additional ribs. Its preservation closely resembles that of the ventral part of the Carboniferous "conodontochordate" *Typhloesus* (Conway Morris, 1990). This ventral structure was more resistant to decay than the body integuments immediately above it. Its walls were rather rigid but elastic, as documented by the observed pattern of deformation, especially in partially decayed specimens. The ventral trough enclosed a large space that probably contained some kind of visceral organs such as the stomach, intestine, or liver.

The head region of *Yunnanozoon* was structurally complex, although no one specimen provides details of its organization. There is a dark-stained structure that is strongly concave and of a ringlike appearance in some specimens. Others show two large lateral lobes with semicircular lateral margins, each probably with holes in the center. They were compared with eye sclerotic rings of the conodonts (Dzik, 1995). Another, ventrally oriented head structure is subdivided into several sclerotic units, probably in excess of 12 (Dzik, 1995), apparently forming together a ring reminiscent of the oral sclerotic ring of the early Silurian anaspid *Jamoytius* (Ritchie, 1968). The high degree of organization in the head of *Yunnanozoon* is unexpected; the possible presence of eyes and a feeding apparatus requires that the neural system was accordingly developed (which is consistent with the reevaluation of neontological data by Wicht and Northcutt, 1992). If the oval head structures in *Yunnanozoon* are truly eyes, this would imply the presence of the neural tube because vertebrate eyes are bulbous extensions of the brain, the photoreceptors being transformed ependymal cells of the brain cavity. Olfactory sense organs apparently preceded eyes in the phylogeny of chordates. This is suggested by their terminal location in the head. Their sensory epithelial cells produce processes to communicate with the brain, a trait definitely very ancient, reminiscent of the innervation mode of muscle cells in the nematodes and in *Amphioxus*.

The body plan of *Yunnanozoon*, although typically chordate, differs substantially from that of Recent *Branchiostoma*. The low number of gill-slits (probably six) appears to be a very old trait, and the presence of an atrium is certainly a derived feature. In having strictly transverse myosepta and robust myomeres, as well as a small number and probably direct openings of particular gill slits, *Yunnanozoon* is more primitive than any known Recent or fossil chordate. It seems, therefore, that *Yunnanozoon* belongs to a completely extinct branch of the earliest chordates, the main characteristics of which are not preserved even in the ontogeny of Recent chordates, being suppressed by later anatomical acquisitions. The embryological evidence does not seem to be matched by paleontological data deeper than to the Middle Cambrian *Pikaia* and *Metaspriggina*, except for the earliest embryonic stages, when the myocoel develops as metamerically arranged vesicles above the long notochord, unlike succeeding somites, and the first is longitudinally elongated. The somites merge and enclose the neural tube in a way similar to the formation of the collar region in enteropneusts.

The only parts of the enteropneust body that may be comparable to the muscular chambers of *Yunnanozoon* is the proboscis and collar (see Benito and Pardos, 1997). There is some resemblance in shape between the first muscular block in *Yunnanozoon* and the proboscis, which probably developed from fused paired cavities. The proboscis coelom in the enteropneusts constitutes an unpaired cavity, but the collar coelom develops by invagination from two lateral pouches and may be separated in two parts by both dorsal and ventral mesenteria. The second pair of muscular chambers in *Yunnanozoon* may be a homolog of the enteropneust collar. The collar neural cord of some enteropneusts has a continuous lumen that opens to the exterior at each end by neuropores. This further supports the possible homology with the muscular chambers of *Yunnanozoon*. Their homology with myomeres of later chordates implies that the nerve cord was located in between them. There is no evidence that enteropneusts were already effective burrowers in the Cambrian; the possibility remains that their proboscis and collar developed from structures more similar to the myomeres of *Yunnanozoon*, present in the common ancestor of both organisms.

The large size of the notochord and the strictly dorsal position of the muscular blocks are difficult to reconcile with presently accepted views of the early phylogeny of the chordates. Yet, there is another Cambrian organism, *Odontogriphus* from the Burgess Shale (Conway Morris, 1976a), that shows a series of metameric transverse units resembling that in *Yunnanozoon* and possibly dorsal in respect to the straight intestine. There is no evidence of a notochord but it had a mouth apparatus that was armed with denticles resembling westergaardodinid conodont elements (paraconodonts).

PARACONODONTS AND THE EVOLUTIONARY ORIGIN OF THE CONODONT APPARATUS

The elements of the oral apparatus of the westergaardodinid conodonts (paraconodonts; Bengtson, 1976) differ both structurally and functionally from typical conodont elements, as their epithelial cover may have disappeared apically at later histogenetic stages. *Odontogriphus* from the Middle Cambrian Burgess Shale (Conway Morris, 1976a) is the only anatomically preserved fossil possibly representing a westergaardodinid. The only known specimen is dorsoventrally compressed, probably corresponding to the original compression of the body (Conway Morris, 1976a). Body segments are very short, with clear, strictly transverse boundaries. Although much wider and dorsoventrally compressed, they may be homologous to the muscular chambers of *Yunnanozoon*. Oval dark structures that border the segments in *Odontogriphus* resemble the gonads in *Yun-*

nanozoon. Lateral unsegmented fields may correspond to lateral fins or to the ventral trough of *Yunnanozoon*. The head region of *Odontogriphus* was not segmented and was equipped with a pair of presumably sensory organs ("palps") of unknown function and homology. The intestine is straight and runs along the midline, perhaps under the segments. Earlier suggested chordate affinities of *Odontogriphus* (Dzik, 1976, 1993) may thus find some support in the anatomy of *Yunnanozoon*. The dorsoventrally compressed oval body of this Middle Cambrian organism may be a trait connecting it with some fossils even older and more primitive than *Yunnanozoon*. The mouth of *Odontogriphus* was armed with a bilobate apparatus, apparently ventral in position. Elements of this apparatus are mineralogically altered, making affinities with the westergaardodinids uncertain, but their external shapes and organization of the apparatus fit expectations.

The internal structure of westergaardodinid elements shows that they originated as organic rods within the "dental papilla" of a developing conodont element (Müller and Nogami, 1971; Szaniawski, 1971) with distinct episodes of organic matrix and mineral secretion. Later in histogeny, secretion was restricted to the base of the elements and as a result, a conical structure gradually developed basally (Müller and Nogami, 1971; Szaniawski, 1971). Secretion ceased at the apex of the element and it may have been exposed at this stage. The growth of the element continued to be stadial (possibly in connection with a circadian rhythm of feeding) with concentric wrinkles visible on both external and internal surfaces of the base. In some genera (*Problematoconites*) a special kind of tissue developed later in histogeny within the basal cavity, which may extend to the outside (Fig. 2a; Andres, 1988). This tissue is perforated by pores and its surface within the basal cavity is irregular, with wrinkles, ribs, and, perforations. It is thus very different from the regular element tissue of the paraconodonts. Only some elements in samples show the presence of this kind of tissue, others are of typically paraconodont organization (structurally not differentiated type of Andres, 1988).

The typical panderodontid or protopanderodontid conodonts (euconodonts of Bengtson, 1976) co-occurred with the westergaardodinids in the Late Cambrian and they are almost certainly phylogenetically related. The most primitive of the euconodonts (*Proconodontus*; Andres, 1988; Fig. 2b, this chapter) have a basal perforated tissue, structurally indistinguishable from that of the paraconodont *Problematoconites*. Szaniawski and Bengtson (1993: Fig. 6: 2), in their schematic drawing of the *Proconodontus* element section, proposed that the basal filling tissue had the same internal structure as the typical westergaardodinid element. This is used to support their hypothesis that the crown tissue of euconodonts was a new acquisition and that the basal filling tissue of euconodonts was homologous with the whole paraconodont element. However, I have not been able to find any convincing evidence of continuity of growth increments between the basal filling tissue and the crown tissue in Andres' (1988) data. To the contrary, the boundary between the basal filling and the crown tissue appears distinct both internally (see Andres, 1988: pl. 10–13) and at the surface of elements (e.g., Chen and Gong, 1986: pl. 23:, 16, 25: 6, 32:, 15), in *Problematoconites*, in *Proconodontus*, and in later conodonts. In all well-known conodont elements, the basal filling tissue and the crown tissue are structurally distinct. The boundary between them is clear-cut and serrate in cross-section. This closely resembles relationships between ganoine and underlying ectomesenchymal tissues in actinopterygian fish scales. Usually, the conodont basal filling tissue was not mineralized and is missing in fossil specimens.

The enigmatic sclerites of *Fomitchella* are composed of a fully mineralized crown tissue that is of coniform morphology from the beginning of its histogeny. Its appearance in the earliest Cambrian (Bengtson, 1983) makes the primitive nature of the paraconodont element organization somewhat uncertain. The possibility remains that the early Cambrian ancestors of the westergaardodinids could have had elements of euconodont organization, thus identical with *Fomitchella*. A secondarily delayed mineralization of the organic matrix of the crown tissue, so common among higher chordates, could have resulted in developing the peculiar paraconodont histogeny.

There are thus at least three possible scenarios for the evolutionary transitions between paraconodonts and euconodonts.

1. The basal filling tissue of the euconodonts originated by developing a clear-cut difference between the secretive properties of epithelial cells covering the paraconodont element from outside and those within the basal cavity (Bengtson, 1976).
2. The inner part of the paraconodont element developed a porosity and transformed into the basal filling tissue and a secretion of strongly mineralized crown tissue was subsequently initiated from outside, whereas the external secretion of the paraconodont element tissue ceased completely (Szaniawski and Bengtson, 1993).
3. The crown tissue of the euconodont is homologous with the whole paraconodont, being different mostly in timing of calcification in the histogeny, whereas the basal filling tissue was generally not mineralized in the conodonts, and if mineralized then it was generally late in histogeny or even diagenetically (Dzik, 1976, 1986). The

periodicity in secretion of the paraconodont tissue may have
resulted from sequential secretion of an organic matrix and its
calcification in a mode similar to the ganoine of Recent fishes.

The homology of the paraconodont element tissue with the ectomesenchymally secreted basal filling tissue of typical conodonts, as proposed by
Bengtson (1976), implies that the initial paraconodont rods were secreted
within a sac built of condensed mesenchyme. This deserves some consideration. In Recent vertebrate teeth a reciprocal signaling between epithelial
and mesenchymal tissues controls the morphogenesis (Thesleff *et al.*, 1995).
Odontoblasts start their secretive activity before ameloblasts and this may
correspond in a way to the proposedly earlier evolutionary origin of the
ectomesenchymal mineral tissue in the conodont lineage. Both the stadial
mode of secretion and mineralization of the paraconodont element and its
morphology are very remote to all known kinds of ectomesenchymal teeth
or scale tissues. This, as well as generally weak and late mineralization of
the undoubtedly ectomesenchymal basal filling tissue, makes this interpretation difficult to accept. Another interpretation is more appealing to me.
The early histogeny of the paraconodont element may be a modified
recapitulation of its evolutionary origin. The organic denticle developing
inside an epithelial pocket and later puncturing it may have originated
from a hook deeply imbedded within the epithelium, in a way similar to
nemertinean stylets. This may explain why such epithelial pockets developed in chordates. Presumably both in histogeny and evolution they
subsequently enveloped the ectomesenchymal dental papilla, in effect
developing a conical shape. The ability to secrete phosphatic tissues seems
to be very ancient in the Metazoa. In the lingulide brachiopods, which are
"living fossils" presumably preserving now the physiologic mechanisms
typical of early Paleozoic organisms, the secretive cells are in direct contact
with the mineralized calcium phosphatic tissue (Williams *et al.*, 1994), a relationship analogous to that in the conodont crown and vertebrate enamel.
According to Williams and colleagues (1994) apatite granules are exocy-
tosed in brachiopods.

This scenario of the origin of the conodont apparatus also has implications for the ancestry of the chordates. If one accepts the (undoubtedly
controversial) idea that *Yunnanozoon* and *Odontogriphus* are the most
primitive chordates, some expectations can be also formulated regarding
the anatomy of the common ancestor of these organisms. It should bear
metamerically arranged muscular chambers with internal cavities separated
from each other by transverse myosepta. A cylindrical intestine, with several
branchial slits or pouches in the anterior part, should run below this
metameric vesicular unit. The notochord may have been an evolutionary

novelty connected with the chordate style of locomotion (by lateral undulations of the body) not necessarily present in *Odontogriphus* and its common ancestor with *Yunnanozoon*. The fossil organism that may appear crucial in these considerations is the Vendian *Dickinsonia*, the only known animal of this geological age with dorsally located metameric muscular blocks and intestinal caeca that could have given rise to both hepatic sacculation of the enteropneusts and branchial pouches.

DICKINSONIA AND THE ANCESTRY OF CHORDATES

The internal anatomy of *Dickinsonia* remains a matter of dispute and countless interpretations of its relationships have already been offered. Nevertheless, despite its preservation on the bedding planes of coarse sandstone, there is surprisingly rich evidence to interpret. Thus, Seilacher (1989) has convincingly shown that particular "segments" of the body were actually chambers separated from each other by walls and filled with a liquid under pressure during life. In the "quilted pneu structure" of *Dickinsonia* there is an anterior medially elongated unpaired unit. At early ontogenetic stages with a low number of muscular units the anteriormost one was much more elongated and rounded triangular in outline (Runnegar, 1982). Behind, it is followed by transversely elongated modules that seem to be at least subdivided medially by a kind of mesenterium in the center of the body (according to Gehling, 1991, only from one side). The *Dickinsonia* body increased in size by adding new metameric units at its posterior end. Wade (1972) and Runnegar (1980) have pointed out concentric wrinkles and changes in diameter of the body in some specimens that can be explained as a muscular contraction (desiccation being a less likely alternative), with muscle fibers running longitudinally (parallel to the body margin: Runnegar, 1982: Fig. 1e). The ventral and dorsal sides of the quilted structure of *Dickinsonia* were apparently of similar morphology (but see Gehling, 1991); so it can hardly be compared with any complete known organism. Internal structures other than the dorsal metameric quilt are preserved only in a few specimens of *Dickinsonia* among several hundreds collected (Wade, 1972, p. 175). However, some specimens show a presence of sediment-filled cylindrical gut under the quilted structure (Runnegar 1982: Fig. 1c; Jenkins, 1985, 1992: Fig. 14). This is the relationship between muscular blocks and the alimentary tract as in *Yunnanozoon* and, proposedly, in *Odontogriphus*, the main difference being that in adult *Dickinsonia* the body was very strongly dorsoventrally flattened.

If the dorsal segmented quilt of *Dickinsonia* is truly homologous to the muscular blocks of *Yunnanozoon* (Dzik, 1995), the polarity of relationship between the hemichordates and the earliest chordates should be the reverse to that generally assumed. In this case, the origin of branchial slits would be phylogenetically later than the appearance of dorsal muscular coelomic pouches. The enteropneusts and pterobranchs would appear to be derived successors of a *Yunnanozoon*-like ancestor of both hemichordates and chordates, although they undoubtedly preserved several very ancient traits in their anatomy, for instance an organization of cilia closely resembling *Xenoturbella* (Benito and Pardos, 1997, p. 24), the most primitive Recent flatworm. This does not significantly change the pattern of relationships and is not contradicted by other kinds of phylogenetic evidence. Sequence data for 18S rRNA suggests that the pterobranchs are closely related to both the enteropneusts and the echinoderms, being not far from the tunicates, whereas the acraniates cluster together with the vertebrates (Turbeville *et al.*, 1994; Halanych, 1995; the proximity of lampreys to myxinoids is also supported—Stock and Whitt, 1992). This is consistent with the proposed derivation of *Amphioxus* (and tunicates) from anatomically rather advanced extinct chordates (good swimmers—with V-shaped myomeres) by secondary anatomical simplification and of enteropneusts from *Yunnanozoon*-like earliest chordates (Dzik, 1995; a somewhat less radical interpretation is shown on Fig. 13). In fact, Halanych (1995, p. 72) suggested that "ciliated gill slits and the dorsal hollow nerve chord are plesiomorphic features of the Deuterostomia." This would indicate that the stomochord of the enteropneusts is a rudiment of the earlier functional (in *Yunnanozoon*) notochord and the simplicity of the nervous system in the pterobranchs is a neotenous feature, related to their small adult size (Dilly, 1975, p. 11). Obviously, all those extinct animals were much more primitive than any Recent craniates, particularly in having only one *Hox* gene cluster, as in *Amphioxus* (see Garcia-Fernàndez and Holland, 1994).

In any case, to explain the origin of the diagnostic anatomical features of *Odontogriphus*, *Yunnanozoon*, and conodonts as interpreted herein one has to invent a hypothetical organism similar to *Dickinsonia*. Its proposed anatomical organization could easily have produced branchial pores by connecting caeca with the body wall and the segmented dorsal quilt may have been transformed into myomeres. The peculiar mode of innervation of muscular blocks in the most primitive chordates by processes of muscle cells, inherited after ancestral worms and preserved in *Amphioxus*, requires that the main neural cord is in the proximity of the locomotory muscles. The original dorsal position of these blocks might have been the main reason for its dorsal position in the chordates.

Much more difficult to answer is the question of the origin of body organization in *Dickinsonia* and its relationship to other phyla. Its anatomy is reminiscent of some features of the nemerteans, particularly its metameric caeca that lack bounding mesenteria (Glaessner and Wade, 1966: pl. 101: 4; Long, 1995: p. 15; here Fig. 11). The presence of metameric gonads in *Odontogriphus* and in *Yunnanozoon* (Chen *et al.*, 1995; Shu, Zhang, and Chen, 1996) is a trait shared not only with the nemerteans but also with the enteropneusts, which makes it a likely primitive (plesiomorphic) feature of these groups. In *Yunnanozoon*, each gonad probably opened separately at the body surface and between branchial slits that may have developed from serial intestinal caeca (if so, this is exactly as in the nemerteans). Some enteropneusts have serial intestinal caeca (hepatic sacculation) of digestive function (Benito and Pardos, 1997). Jenkins (1992, p. 163, Fig. 13) restored a dorsally located longitudinal intestinal diverticulum in *Dickinsonia*, which may correspond to the notochord or the rhynchocoel. An oral apparatus had no chance of having been preserved in the Ediacaran specimens of *Dickinsonia*, but in Recent nemerteans there are sclerites with some resemblance to paraconodonts (the group to which *Odontogriphus* possibly belongs). The proboscis stylets of the nemerteans originate in epithelium (inside enlarged cells) and are of a shape similar to that of the earliest stages in development of the westergaardodinid elements, being composed of a central organic matrix surrounded by an inorganic cortex containing calcium phosphate (Stricker, 1982). It is generally accepted that the presence of numerous stylets is primitive for the Nemertini (Gibson, 1988; Crandall, 1993). The nemertean features presumably common to the *Dickinsonia*, *Odontogriphus*, and *Yunnanozoon* anatomies recall the old idea of relationships between the Nemertini and Chordata (Jensen, 1960, 1988; E. N. Willmer, 1974, 1975). Some similarities in the excretory and reproductive systems of the nemerteans and chordates have already been pointed out by E. N. Willmer (1974). Perhaps these three Precambrian–Cambrian organisms represent a link connecting these two phyla. The evidence is obviously very weak, as usual in paleontology, but it calls for more attention to be paid to this hypothesis, now considered rather unorthodox. An additional argument in its favor may be the inability of the nemertineans to secrete chitin, shared by them with sipunculans, hemichordates, echinoderms, and chordates (P. Willmer, 1990, p. 80). Molecular phylogenies, however, place nemertineans far from the chordates (Turbeville *et al.*, 1992; Winnepenninckx *et al.*, 1995).

If the Middle Cambrian *Amiskwia* (Fig. 12*a*) belongs to the Pelagonemertini (Conway Morris, 1977, did not find any evidence for a retractable proboscis), they were highly advanced anatomically already in the Cambrian. Recent pelagic nemerteans show several primitive features in their

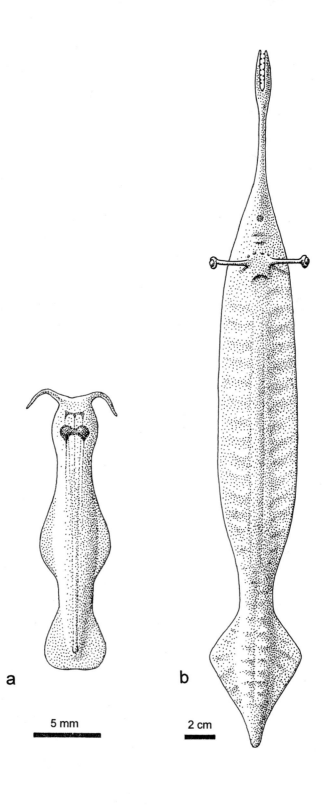

a

5 mm

b

2 cm

anatomical organization (Crandall, 1993) and may be a relic group. The enigmatic Carboniferous *Tullimonstrum* (despite various interpretations of its affinities; see e.g., Beall, 1991, and Bousfield, 1995), with its internal metameric body, stalked eyes, and grasping apparatus at the end of its long proboscis (Fig. 12*b*) shows a combination of characters that may express a remote relationship with a nemertean ancestor of the same stock. Most interestingly, the proboscis of *Tullimonstrum* was not retractable but, instead, it was bilobed and armed with scleritized, possibly mineralized denticles (Foster, 1979). The rhynchocoel in which the nemertean proboscis is retracted bears some resemblance to the anterior medial chamber of the quilt of *Dickinsonia* and to the enteropneust proboscis. The origin of the retractable proboscis apparatus could have been preceded by an extension of the anterior segment of the body together with the perioral apparatus (the mouth opening keeping its original position) and *Tullimonstrum* may have preserved to the Carboniferous some features of such a transitional stage. The origin of nemerteans from anatomically more complex coelomates has been already forwarded by Turbeville and Ruprecht (1985).

Possible nemertean ancestry of the chordates is in conflict not only with the traditional view that the anatomical organization of hemichordates, highly ecologically specialized in two opposite directions (tentacle filtration in clonal pterobranchs and mud burrowing in enteropneusts), represents a connecting link between the invertebrates and the chordates. It is also in apparent conflict with the idea that conodont elements originated from grasping spines of the Cambrian chaetognaths ("protoconodonts").

THE PROBLEM OF PROTOCONODONTS

Although the pocket theory of conodont elements functioning is apparently falsified by the microornamental features of the conodont element surface, the idea of protoconodont → paraconodont → euconodont

FIG. 12. Problematic Paleozoic animals of possible nemertean affinities: (a) *Amiskwia sagittiformis*, from the Middle Cambrian Burgess Shale of British Columbia (after Conway Morris 1977); (b) *Tullimonstrum gregarium* from the Late Carboniferous Mazon Creek fauna of Illinois (based on data of Foster, 1979, and Beall, 1991).

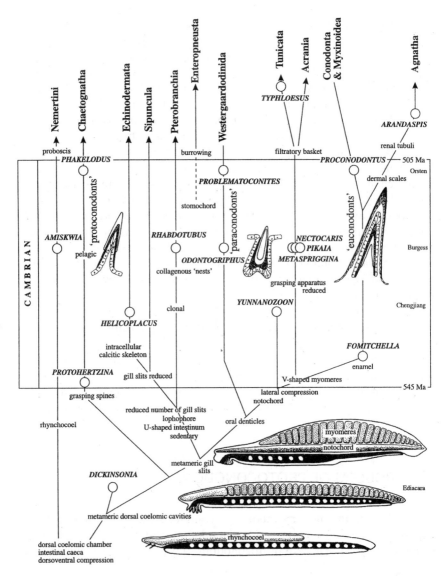

FIG. 13. The most parsimonious "chronophyletic" (see Dzik, 1991*b* for a discussion of methodology) scenario of evolutionary relationships of the chaetognaths ("protoconodonts") and the conodonts ("paraconodonts" and "euconodonts"; diagrammatic sections show their mode of secretion) in respect to other early metazoans. The alternative would be that the grasping apparatuses of the chaetognaths and conodonts are homologous (Bengtson, 1976), which implies that the anatomy of Recent chaetognaths is secondarily simplified (Christofferson and Araújo-de-Almeida, 1994) and they restored ability to secrete chitin; the tail of the probable paraconodont *Odontogriphus* should then be laterally compressed (the specimen is too poorly known to exclude or to prove this) to make it anatomically closer to the conodonts. Another possibility is that the chaetognaths, with their externally secreted chitinous grasping spines, originated from ancestors at the nemertean grade before they lost chitin synthetase and developed internally secreted stylets.

relationship (Bengtson, 1976) does not necessarily need to be connected with the theory and requires a separate discussion. The earlier step in this speculative evolutionary transition is confronted in the following paragraphs with the available evidence and inference based on it.

The protoconodonts are narrowly conical sclerites, known to form grasping apparatuses virtually indistinguishable from those of the Recent chaetognaths and with basically the same internal structure (Szaniawski, 1982). Like the chaetognath grasping spines (built of α-chitin; Kapp, 1991), they were secreted at the surface of long conical soft tissue appendages. The idea of the chaetognath nature of protoconodonts is the most parsimonious solution of the problem of their affinities and as such should be accepted, although it has to wait for corroboration by anatomical evidence. No fossil documents an association of the protoconodonts with any soft parts. It remains possible, even if the protoconodont animals were truly ancestors of the Chaetognatha, that they were anatomically very different from their living relatives. This is suggested by the occurrences of the oldest known protoconodonts of the earliest Cambrian in relatively shallow-water sediments, unlike their Late Cambrian successors, which were generally confined to pelagic black shales and limestones.

The paraconodonts (elements of the westergaardodinid conodonts) are widely conical in shape and their internal structure is basically different from that of the protoconodonts. However profound are the structural differences between the proto-, the para-, and the euconodonts, there is nothing in their structure that would make evolutionary connections between them impossible. The real problem is with the anatomy of their bearers. The most reasonable interpretation of the protoconodonts as chaetognath grasping spines implies that the Cambrian *Protohertzina* or *Phakelodus* were unsegmented animals with bodies stiffened by coelomic fluid pressure and with lateral fins propelling them by dorsoventral undulations of the tail. The mouth is terminal in chaetognaths, the anus being located ventrally, with a tail behind it. In contrast, the conodonts were true chordates with a laterally compressed body stiffened by the internally located notochord and propelled by lateral waving of the tail. Transversely straight segments and a dorsoventrally flattened body characterize the possible westergaardodinid *Odontogriphus* (Conway Morris, 1976a). It appears thus that grouping together proto-, para-, and euconodonts, based on structural similarities of their oral sclerites would encompass anatomies of at least two, if not three, phyla. This is, again, not impossible but any hypothesis of homology between these skeletal elements would require that the common ancestor of all these organisms possessed a grasping oral apparatus (Fig. 13).

The inferred presence of an organic cap above the mineralized portion of the conodont element invokes a question of whether the chaetognath organic grasping tissue is not homologous with this protective structure. The crown tissue of the conodont element would then develop secondarily within the invaginated internal epidermis of the papilla. Such a model does not encompass an explanation of the origin of the ectodermal enamel organ. This is more easily resolved if embedding of an ectodermally derived denticle (stylet) and its subsequent overgrowth by the epithelium is assumed. Any direct relationship between the chaetognaths and the chordates requires that the low number of coelomic compartments in the Chaetognatha was a result of secondary reduction in the number of originally metameric chambers (Fig. 13). In fact, Christoffersen and Araujo-de-Almeida (1994) have proposed that the chaetognaths secondarily lost the gill slits, the notochord, and the endostyle although they do not provide any evidence for this and their ideas are apparently influenced by Bengtson's (1983) interpretation of conodont–protoconodont relationships. Some features of chaetognath anatomy may perhaps be explained in this way, for instance the inverted photoreceptor cells in eyes of most chaetognaths and a multilayered sheath reminiscent of vertebrate myelin, which covers their cerebral ganglion (Bone and Goto, 1991; but nerve fibers in *Myxine* and other agnathans are unmyelinated, with only some overlap of Schwann cells sheath, Peters, 1963). The transverse muscle cells that occur in several chaetognaths, regarded as the more primitive, penetrate the basement membrane to reach nerve terminals in the epidermis (Bone and Duvert, 1991). This is reminiscent of the situation in *Amphioxus* and in nematodes, where muscle cells produce processes that terminate at nerve cords (Wright, 1991). In fact, according to Nielsen and Colleagues (1996, p. 401), the early embryogeny of the chaetognaths is typical of aschelminthes and the chitinous oral teeth closely resemble the mastax of the rotifers. This would make the chaetognaths an early offshoot of the priapulids but does not contradict a distant relationship to the chordates (although molecular data suggest that the chaetognaths are distant from the deuterostome; Wada and Satoh, 1994). It has to be stressed, however, that there is a contradiction between the possible homology of the organic cap of the conodont elements with the teeth of *Myxine* and a suggestion of homology with the grasping spines of the chaetognaths.

CONCLUSIONS

There are several anatomical traits shared by the most primitive Early Cambrian chordates *Yunnanozoon*, the Middle Cambrian probable westergaardodinid *Odontogriphus*, and the Vendian problematicum *Dickinso-*

nia, and their body plans resemble that of the nemerteans. The rhynchocoel of the nemerteans and (at least) the most anterior dorsal chamber of *Dickinsonia* may appear thus homologous, which is consistent with the idea of its homology with the notochord and myocoel of the chordates. The phosphatic elements of the westergaardodinid paraconodont apparatuses seem to represent a stage in the development of mineral skeleton transitional between nemertean stylets and conodont elements. *Odontogriphus* had an oral grasping apparatus composed of such mineralized, probably phosphatic elements. Some kind of oral apparatus of unknown internal structure also occurred in *Yunnanozoon*, which probably also bore complex sensory head organs. *Dickinsonia, Odontogriphus, Yunnanozoon*, and *Pikaia* may thus represent a developmental series, from muscular dorsal chambers to the organization of myomeres typical for all later chordates (Fig. 13). In this respect even cephalochordates and tunicates are anatomically highly derived. *Amphioxus* and the tunicates share the presence of the filtratory basket surrounded by the atrium, which indicates that they are closely related and evolutionarily late. The earliest, more or less reliable fossil evidence of this structure is in the Carboniferous *Typhloesus*. It shares with Recent salps a very characteristic concentration of the alimentary tract in a globular body (although bilobed in *Typhloesus*). Notably, the latest of the anaspid agnathans with weakly mineralized body covers, the Late Devonian *Legendrolepis*, shows a very elaborate gill apparatus (Arsenault and Janvier, 1991), suggestive of filtratory adaptations. The time order in appearance of these anatomies is consistent with the idea that salps and *Amphioxus* derived from chordates related to the anaspids. This implies a more complex anatomy of the common ancestor of the tunicate and cephalochordate evolutionary branches than is usually assumed, but it is consistent with neontological data by Lacalli (1996) and Williams and Holland (1996).

The mode of secretion of the conodont crown tissue, with its early mineralization, and distribution of shapes of the secretory cells suggest that localized intensity of secretion of calcium phosphate, instead of cell migration, was the main factor controlling the morphology of elements. Calcium phosphate secretion was extremely high at denticle tips, which requires (as phosphate and calcium ion transport took place above the surface of the element) a relatively thick cover of soft tissue to fulfill supply needs. In typical conodonts, this tissue was probably protected on the outside with horny caps. The primitive panderodontid conodonts had a grasping apparatus composed of at least seven element pairs morphologically similar to that of the Cambrian and Recent chaetognaths. Any homology of their crown tissue or the organic cap, with the grasping spines of the chaetognaths (protoconodonts), is unlikely. The typical organization of the conodont apparatus originated in the early Ordovician within the clade of coniform Protopanderodontida. Their apparatuses were composed of a pair of incisorlike elements in front, a set of four pairs of relatively gracile elements connected into a single unit by one posteriorly located symmetrical element, and two pairs of robust elements hidden within the throat. This apparatus architecture did not undergo any basic modifications in the ozarkodinid and prioniodontid clades. In the most advanced prioniodontids (e.g., *Promissum*) the apparatus was probably evertible, with the throat being sinuously folded at the resting position. Elements of some locations were doubled in these conodonts. A homology between the organic caps of elements of advanced conodonts and horny jaws of Recent hagfish is likely, which would make the Myxinoidea close relatives of the conodonts or even members of their class.

The homology between the crown and basal filling tissues of the conodonts and the enamel and dentine of vertebrates, respectively, implies that the developmental mechanisms that arose at the origin of the conodonts were subsequently used by the agnathans to build up their protective dermal scales.

ACKNOWLEDGMENTS

I am very thankful to Philip C. J. Donoghue (University of Birmingham) for his comments on the manuscript and for improving its language and style. I also deeply appreciate constructive criticism offered by Richard J. Aldridge (University of Leicester) in his review, which resulted in a substantial improvement of the text. I am especially grateful for his suggestions on how to modify the diagram of relationships. I also appreciate helpful comments and suggestions by an anonymous reviewer.

REFERENCES

Aldridge, R. J., 1982, A fused cluster of coniform elements from the late Ordovician of Washington Land, western North Greenland, *Palaeontology* **25:**425–430.
Aldridge, R. J., Briggs, D. E. G., Clarkson, E. N. K., and Smith, M. P., 1986, The affinities of conodonts—New evidence from the Carboniferous of Edinburgh, Scotland, *Lethaia* **19:**279–291.

Aldridge, R. J., Briggs, D. E. G., Smith, M. P., Clarkson, E. N. K., and Clark, N. D. L., 1993, The anatomy of conodonts, *Philosophical Transactions of the Royal Society London B* **340:**405–421.

Aldridge, R. J., Purnell, M. A., Gabbott, S. E., and Theron, J. N., 1995, The apparatus architecture and function of *Promissum pulchrum* Kovács-Endrödy (Conodonta, Upper Ordovician) and the prioniodontid plan, *Philosophical Transactions of the Royal Society London B* **347:**275–291.

Aldridge, R. J., Smith, M. P., Norby, R. D., and Briggs, D. E. G., 1987, The architecture and function of Carboniferous polygnathacean conodont apparatuses, in: *Palaeobiology of Conodonts*, (R. J. Aldridge, Ed.) pp. 77–90, British Micropalaeontological Society Series.

Aldridge, R. J., and Theron, J. N., 1993, Conodonts with preserved soft tissue from a new Ordovician *Konservat Lagerstätte, Journal of Micropalaeontology* **12:**113–117.

Andres, D., 1988, Strukturen, Apparate und Phylogenie primitiver Conodonten, *Palaeontographica A* **200:**105–152.

Arsenault, M., and Janvier, P., 1991, The anaspid-like craniates of the Escuminac Formation (Upper Devonian) from Miguasha (Quebec, Canada), with remarks on anaspid-petromyzontid relationships, in: *Early Vertebrates and Related Problems of Evolutionary Biology* (M. M. Chang and G. R. Zhang, Eds.), pp. 19–40, Science Press, Beijing.

Bard, J. 1990, Morphogenesis; The Cellular and Evolutionary Process of Developmental Anatomy, Cambridge University Press, Cambridge.

Bardack, D., and Richardson, E. S., Jr., 1977, New agnathous fishes from the Pennsylvanian of Illinois, *Fieldiana Geology* **33:**489–510.

Beall, B. S., 1991, The Tully monster and a new approach to analyzing problematica, in: *The Early Evolution of Metazoa and the Significance of Problematic Taxa* (A. M. Simonetta and S. Conway Morris Eds.), pp. 271–285, Cambridge University Press, Cambridge.

Bengtson, S., 1976, The structure of some Middle Cambrian conodonts, and the early evolution of conodont structure and function, *Lethaia* **9:**185–206.

Bengtson, S., 1983, The early history of the Conodonta, *Fossils and Strata* **15:**5–19.

Benito, J., and Pardos, F., 1997, Hemichordata, in: *Microscopic Anatomy of Invertebrates, Vol. 15, Hemichordata, Chaetognatha, and the Invertebrate Chordates* (F. W. Harrison and E. E. Ruppert, Eds.), pp. 15–101. Wiley-Liss, New York.

Bone, Q., and Duvert, M., 1991, Locomotion and buoyancy, in: *The Biology of Chaetognaths* (Q. Bone, H. Kapp and A. C. Pierrot-Bults, Eds.), pp. 32–44. Oxford University Press, Oxford.

Bone, Q., and Goto, T. 1991, The nervous system, in: *The Biology of Chaetognaths* (Q. Bone, H. Kapp and A. C. Pierrot-Bults, Eds.), pp. 18–31. Oxford University Press, Oxford.

Bone, Q., Ryan, K. P., and Pulsford, A. L. 1983, The structure and composition of the teeth and grasping spines of chaetognaths, *Journal of the Marine Biological Association of the United Kingdom* **63:**929–939.

Bousfield, E. L., 1995, A contribution to the natural classification of lower and middle Cambrian arthropods: Food-gathering and feeding mechanisms, *Amphipacifica* **2:**3–34.

Boyde, A., 1978, Development of the structure of the enamel of the incisor teeth in the three classical subordinal groups of the Rodentia, in: *Development, Function and Evolution of Teeth* (P. M. Butler and K. A. Joysey, Eds.), pp. 43–58, Academic Press, London.

Brigger, D., and Muckle, T. J., 1975, Compraison of Sirius Red and Congo Red as stains for amyloid in animal tissues, *The Journal of Histochemistry and Cytochemistry* **23:**84–88.

Briggs, D. E. G., and Clarkson, E. N. K., 1987, An enigmatic chordate from the Lower Carboniferous "shrimp-bed" of the Edinburgh district, Scotland, *Lethaia* **20:**107–115.

Briggs, D. E. G., Clarkson, E. N. K., and Aldridge, R. J., 1983, The conodont animal, *Lethaia* **16**:1–14.

Burnett, R., 1988, Polygonal ornament in the conodont *Siphonodella*: An internal record, *Lethaia* **21**:411–415.

Burnett, R. D., and Hall, J. C., 1992, Significance of ultrastructural features in etched conodonts, *Journal of Paleontology* **66**:266–276.

Carls, P., 1977, Could conodonts be lost and replaced? Numerical relations among disjunct conodont elements of certain Polygnathaceae (late Silurian—Lower Devonian, Europe), *Neues Jahrbuch für Geologie und Paläontologie, Abhandlungen* **155**:18–64.

Chen J.-y., and Gong W.-l., 1986, Conodonts, in: *Aspects of Cambrian-Ordovician Boundary in Dayangcha, China* (Chen Jun-yuan, Ed.), pp. 93–223. China Prospect Publishing House, Bejing.

Chen, J. Y., Dzik, J., Edgecombe, G. D., Ramsköld, L., and Zhou, G.-Q., 1995, A possible Early Cambrian chordate. *Nature* **377**:720–722.

Christofferson, M. L., and Araújo-de-Almeida, E., 1994, A phylogenetic framework of the Enterocoela (Metameria: Coelomata), *Revista Nordestina de Biologia* **9**:173–208.

Conway Morris, S., 1976a, A new Cambrian lophophorate from the Burgess Shale of British Columbia, *Palaeontology* **19**:199–222.

Conway Morris, S., 1976b, *Nectocaris pteryx*, a new organism from the Middle Cambrian Burgess Shale of British Columbia, *Neues Jahrbuch für Geologie und Paläontologie, Monatshefte* **1976**:705–713.

Conway Morris, S., 1977, A redescription of the Middle Cambrian worm *Amiskwia sagittiformis* Walcott from the Burgess Shale of British Columbia, *Paläontologische Zeitschrift* **51**:271–287.

Conway Morris, S., 1990, *Typhloesus wellsi* (Melton & Scott 1973), a bizarre metazoan from the Carboniferous of Montana, *Philosophical Transactions of the Royal Society of London* **B 327**:595–624.

Conway Morris, S., and Harper, E., 1988, Genome size in conodonts (Chordata): Inferred variations during 270 million years, *Science* **241**:1230–1232.

Conway Morris, S., and Whittington, H. B., 1979, The animals of the Burgess Shale. *Scientific American* **241**:122–133.

Crandall, F., 1993, Major characters and enoplan systematics, *Hydrobiologia* **266**:115–140.

Dawson, J. A., 1963, The oral cavity, the "jaws" and the horny teeth of *Myxine glutinosa*, in: *The Biology of Myxine* (A. Brodal and R. Fänge, Eds.), pp. 231–255, Universitetsforlaget, Oslo.

Denham, R. L., 1944, Conodonts, *Journal of Paleontology* **18**:216–218.

Dilly, P. N., 1975, The pterobranch *Rhabdopleura compacta*: Its nervous system and phylogenetic position, *Symposia of the Zoological Society of London* **36**:1–16.

Donoghue, P. C. J., 1998, Growth and patterining in the conodont skeleton, *Philosophical Transactions of the Royal Society London B.* **353**:633–666.

Dzik, J., 1976, Remarks on the evolution of Ordovician conodonts, *Acta Palaeontologica Polonica* **21**:395–455.

Dzik, J., 1986, Chordate affinities of the conodonts, in: *Problematic Fossil Taxa* (A. Hoffman and M. H. Nitecki, Eds.), pp. 240–254, Oxford University Press, New York.

Dzik, J., 1991a, Evolution of oral apparatuses in conodont chordates, *Acta Palaeontologica Polonica* **36**:265–323.

Dzik, J., 1991b, Features of the fossil record of evolution, *Acta Palaeontologica Polonica* **36**:2, 91–113.

Dzik, J., 1993, Early metazoan evolution and its fossil record, *Evolutionary Biology* **27**:339–386.

Dzik, J., 1994, Conodonts of the Mójcza Limestone, in: J. Dzik, E. Olempska and A. Pisera. Ordovician carbonate platform of the Holy Cross Mountains, *Palaeontologia Polonica* **53:**43–128.

Dzik, J., 1995, *Yunnanozoon* and the ancestry of chordates, *Acta Palaeontologica Polonica* **40:**341–360.

Dzik, J., 1996, Organization, function, morphogenesis, and origin of oral apparatuses in the conodont chordates, *Sixth European Conodont Symposium (ECOS VI), Abstracts,* 16.

Dzik, J., 1997, Emergence and succession of Carboniferous conodont and ammonoid communities in the Polish part of the Variscan sea, *Acta Palaeontologica Polonica* **42:**57–170.

Dzik, J., and Drygant, D., 1986, The apparatus of panderodontid conodonts, *Lethaia* **19:**133–141.

Dzik, J., and Trammer, J., 1980, Gradual evolution of conodontophorids in the Polish Triassic, *Acta Palaeontologica Polonica* **25:**55–89.

Fedonkin, M. A., 1987, Non-skeletal fauna of the Vendian and its place in the evolution of Metazoa, *Trudy Paleontologičeskogo Instituta AN SSSR* **226:**1–178 [In Russian].

Fels, L. M., Zanz-Altamira, P. M., Decker, B., Elger, B., and Stolte, H., 1993, Filtration characteristics of the single isolated perfused glomerulus of *Myxine glutinosa, Renal Physiology and Biochemistry* **16:**276–284.

Flood, P. R., 1975, Fine structure of the notochord of *Amphioxus, Symposia of the Zoological Society of London* **36:**81–104.

Forey, P., and Janvier, P., 1993, Agnathans and the origin of jawed vertebrates, *Nature* **361:**129–134.

Foster, M. W., 1979, A reappraisal of *Tullimonstrum gregarium,* in: *Mazon Creek Fossils* (M. H. Nitecki, Ed.), pp. 269–301. Academic Press, New York.

Gabbott, S. E., Aldridge, R. J., and Theron, J. N., 1995, A giant conodont with preserved muscle tissue from the Upper Ordovician of South Africa, *Nature* **374:**800–802.

Garcia-Fernàndez, J., and Holland, P. W. H., 1994, Archetypal organization of the amphioxus *Hox* gene cluster, *Nature* **370:**563–566.

Gee, H., 1996, *Before the Backbone: Views on the Origin of Vertebrates,* Chapman & Hall, London.

Gehling, J. G., 1991, The case for Ediacaran fossil roots to the metazoan tree, *Memoirs of the Geological Society of India* **20:**181–224.

Gibson, R., 1988, Evolutionary relationships between mono- and polystiliferous hoplonemerteans: *Nipponnemertes* (Cratenemertidae), a "missing link" genus? *Hydrobiologia* **156:**61–74.

Glaessner, M. F., and Wade, M., 1966, The late Precambrian fossils from Ediacara, South Australia, *Palaeontology* **9:**599–628.

Gross, W., 1957, Über die Basis der Conodonten, *Paläontologische Zeitschrift* **31:**78–91.

Halanych, K. M., 1995, The phylogenetic position of the pterobranch hemichordates based on 18S rDNA sequence data, *Molecular Phylogenetics and Evolution* **4:**72–76.

Hardisty, M. W., and Rovainen, C. M., 1982, Morphological and functional aspects of the muscular system, in: *The Biology of Lampreys, Vol. 4A* (M. W. Hardisty and I. C. Potter, Eds.), pp. 137–231, Academic Press, New York.

Hass, W. H., 1941, Morphology of conodonts, *Journal of Paleontology* **15:**71–81.

Holland, P. W. H., and Garcia-Fernandez, J., 1996, *HOX* genes and chordate evolution, *Developmental Biology* **173:**382–395.

Holland, P. W. H., Garcia-Fernandez, J., Williams, N. A., and Sidow, A., 1994, Gene duplications and the origins of vertebrate development, *Development, Suppl.:*125–133.

Hou, X., Ramsköld, L., and Bergström, J., 1991, Composition and preservation of the Chengjiang fauna—a Lower Cambrian soft-bodied biota. *Zoologica Scripta* **20:**395–411.

Hu, C. C., Sakakura, Y., Sasano, Y., Shum, L., Bringas, P. Jr., Werb, Z., and Slavkin, H. C., 1992, Endogenous epidermal growth factor regulates the timing and pattern of embryonic mouse molar tooth morphogenesis, *International Journal of Developmental Biology* **36:**505–516.

Jefferies, R. P. S., 1986, *The Ancestry of the Vertebrates*, 376 pp., British Museum (Natural History).

Jenkins, R. J. F., 1992, Functional and ecological aspects of Ediacaran assemblages, in: *Origin and Early Evolution of the Metazoa* (J. H. Lipps and P. W. Signor, Eds.), pp. 131–176, Plenum Press, New York.

Jensen, D. D., 1960, Hoplonemertines, myxinoids and deuterostome origins, *Nature* **188:** 649–650.

Jensen, D. D., 1988, Hubrecht, Macfarlane, Jensen and Willmer: On the nature and testability of four versions of the nemertean theory of vertebrate origins, *Hydrobiologia* **156:**99–104.

Jeppsson, L., 1979, Conodont element function, *Lethaia*, **12:**153–171.

Jernvall, J., Kettunen, P., Karavanova, I., Martin, L. B., and Thesleff, I., 1994, Evidence for the role of enamel knot as a control center in mammalian tooth cusp formation: Nondividing cells express growth stimulation *Fgf-4* gene, *International Journal of Developmental Biology* **38:**463–469.

Kapp, H., 1991, Morphology and anatomy, in: *The Biology of Chaetognaths* (Q. Bone, H. Kapp and A. C. Pierrot-Bults, Eds.), pp. 5–17, Oxford University Press, Oxford.

Kemp, A., and Nicoll, R. S., 1995, Protochordate affinities of conodonts, *Courier Forschungsinstitut Senckenberg* **182:**235–245.

Klapper, G., and Bergström, S. M., 1984, The enigmatic Middle Ordovician fossil *Archaeognathus* and its relation to conodonts and vertebrates, *Journal of Paleontology* **58:**949–976.

Krejsa, R. J., Bringas, P., and Slavkin, H. C., 1990a, A neontological interpretation of conodont elements based on agnathan cyclostome tooth structure, function, and development, *Lethaia* **23:**359–378.

Krejsa, R. J., Bringas, P., and Slavkin, H. C., 1990b, The cyclostome model: An interpretation of conodont element structure and function based on cyclostome tooth morphology, function, and life history, *Courier Forschungsinstitut Šenckenberg* **118:**473–492.

Lacalli, T. C., 1996, Frontal eye circuitry, rostral sensory pathways and brain organization in amphioxus larvae: Evidence from 3D reconstructions, *Philosophical Transactions of the Royal Society London B* **351:**243–263.

Lindström, M., and Ziegler, W., 1971, Feinstrukturelle Untersuchungen an Conodonten. 1. Die Überfamilie Panderodontacea, *Geologica et Palaeontologica* **5:**9–33.

Long, J. A., 1995, *The Rise of Fishes. 500 million Years of Evolution*, 223 pp., The John Hopkins University Press, Baltimore.

MacKenzie, A., Ferguson, M. W., and Sharpe, P. T., 1992, Expression patterns of the homeobox gene, *Hox-8*, in the mouse embryo suggest a role in specifying tooth initiation and shape, *Development* **115:**403–420.

Mallatt, J., 1996, Ventilation and the origin of jawed vertebrates: A new mouth, *Zoological Journal of the Linnean Society* **117:**329–404.

Männik, P., and Aldridge, R. J., 1989, Evolution, taxonomy, and relationships of the Silurian conodont *Pterospathodus*, *Palaeontology* **32:**893–906.

Marotta, M., and Martino, G., 1985, Sensitive spectrophotometric method for the quantitative estimation of collagen. *Analytical Biochemistry* **150:**86–90.

Marsal, D., and Lindström, M., 1972, A contribution to the taxonomy of conodonts: The statistical reconstruction of fragmented populations, *Geologica et Palaeontologica SB* **1:**43–46.

Masuda, T., Nishikawa, K., and Takagi, T., 1989, Ultrastructure of secretory ameloblasts in the house musk shrew, *Suncus murinus*, Insectivora, *Acta Anatomica, Basel* **134:**72–78.

Matsuo, S., Ichikawa, H., Wakisaka, S., and Akai, M., 1992, Changes of cytochemical properties in the Golgi apparatus during *in vivo* differentiation of the ameloblast in developing rat molar tooth germs, *Anatomical Records* **234:**469–478.

Meunier, F. J., 1980, Recherches histologiques sur le squelette dermique des Polypteridae, *Archives de Zoologie Expérimentale et Générale* **121:**279–295.

Müller, K. J., and Nogami, Y., 1971, Über den Feinbau der Conodonten, *Memoirs of the Faculty of Sciences, Kyoto University, Series of Geology and Mineralogy* **38:**1–87.

Nicoll, R. S., 1995, Conodont element morphology, apparatus reconstructions and element function: A new interpretation of conodont biology with taxonomic implications, *Courier Forschungsinstitut Senckenberg* **182:**247–262.

Nielsen, C., Scharff, N., and Eibye-Jacobsen, D., 1996, Cladistic analysis of the animal kingdom, *Biological Journal of the Linnean Society* **57:**385–410.

Partanen, A. M., and Thesleff, I., 1989, Growth factors and tooth development, *International Journal of Developmental Biology* **33:**165–172.

Peters, A., 1963, The peripheral nervous system, in: *The Biology of Myxine* (A. Brodal and R. Fänge, Eds.), pp. 92–123, Universitetsforlaget, Oslo.

Pierce, R. W., and Langenheim, R. L., 1970, Surface pattern on selected Mississippian conodonts, *Geological Society of America Bulletin* **81:**3225–3236.

Priddle, J., 1974, The function of conodonts, *Geological Magazine* **111:**255–257.

Pridmore, P. A., Barwick, R. E., and Nicoll, R. S., 1996, Soft anatomy and the affinities of conodonts. *Lethaia* **29:**325–328.

Prostak, K. S., Seifert, P., and Skobe, Z., 1993, Enameloid formation in two tetraodontiform fish species with high and low fluoride contents in enameloid, *Archive of Oral Biology* **38:**1031–1044.

Purnell, M. A., 1993, The *Kladognathus* apparatus (Conodonta, Carboniferous): Homologies with ozarkodinids, and the prioniodinid Bauplan, *Journal of Paleontology* **67:**875–882.

Purnell, M. A., 1994, Skeletal ontogeny and feeding mechanisms in conodonts, *Lethaia* **27:**129–138.

Purnell, M. A., 1995b, Microwear on conodont elements and macrophagy in the first vertebrates, *Nature* **374:**798–800.

Purnell, M. A., Aldridge, R. J., Donoghue, P. C. J., and Gabbott, S. E., 1995, Conodonts and the first vertebrates, *Endavour* **19:**20–27.

Purnell, M. A., and Donoghue, P. C. J., 1997, Architecture and functional morphology of the skeletal apparatus of ozarkodinid conodonts. *Philosophical Transactions of the Royal Society London B.* **352:**1545–1564.

Ritchie, A., 1968, New evidence on *Jamoytius kerwoodi* White, an important ostracoderm from the Silurian of Lanarkshire, Scotland, *Palaeontology* **11:**21–39.

Runnegar, B., 1982, Oxygen requirements, biology and phylogenetic significance of the late Precambrian worm *Dickinsonia*, and the evolution of the burrowing habit, *Alcheringa* **6:**223–239.

Ruppert, E. E., 1997, Cephalochordata (Acrania), in: *Microscopic Anatomy of Invertebrates, Vol. 15, Hemichordata, Chaetognatha, and the Invertebrate Chordates* (F. W. Harrison and E. E. Ruppert, Eds.), pp. 349–504, Wiley-Liss, New York.

Sansom, I. J., 1996, *Pseudooneotodus*: A histological study of an Ordovician to Devonian vertebrate lineage, *Zoological Journal of the Linnean Society* **118:**47–57.

Sansom, I. J., Armstrong, H. A., and Smith, M. P., 1995, The apparatus architecture of *Panderodus* and its implications for coniform conodont classification, *Palaeontology* **37:**781–799.

Sansom, I. J., Smith, M. P., Armstrong, H. A., and Smith, M. M., 1992, Presence of the earliest vertebrate hard tissues in conodonts, *Science* **256:**1308–1311.

Sansom, I. J., Smith, M. P., and Smith, M. M., 1994, Dentine in conodonts, *Nature* **368:**591.

Sansom, I. J., Smith, M. M., and Smith, M. P., 1996, Scales of thelodont and shark-like fishes from the Ordovician of Colorado, *Nature* **379:**628–630.

Schmidt, H., and Müller, K. J., 1964, Weitere Funde von Conodonten-Gruppen aus dem oberen Karbon des Sauerlandes, *Paläontologische Zeitschrift* **38:**105–135.

Seilacher, A., 1989, Vendozoa: Organismic construction in the Proterozoic biosphere, *Lethaia* **22:**229–239.

Shellis, R. P., 1978, The role of the inner dental epithelium in the formation of the teeth in fish, in: *Development, Function and Evolution of Teeth* (P. M. Butler and K. A. Joysey, Eds.), pp. 31–42, Academic Press, London.

Shu, D.-G., Conway Morris, S., and Zhang, X.-L., 1996, A *Pikaia*-like chordate from the Lower Cambrian of China, *Nature* **384:**157–158.

Shu, D.-G., Zhang, X.-L., and Chen, L., 1996, Reinterpretation of *Yunnanozoon* as the earliest known hemichordate, *Nature* **380:**428–430.

Simonetta, A. M., and Insom, E., 1993, New animals from the Burgess Shale (Middle Cambrian) and their possible significance for the understanding of the Bilateria, *Bolletino Zoologico* **60:**97–107.

Sire, J.-Y., 1994, Light and TEM study on nonregenerated and experimentally regenerated scales of *Lepisosteus oculatus* (Holostei) with particular attention to ganoine formation, *The Anatomical Record* **240:**189–207.

Sire, J.-Y., 1995, Ganoine formation in the scales of primitive actinopterygian fishes, lepisosteids and polypterids, *Connective Tissue Research* **33:**213–222.

Sire, J. Y., Géraudie, J., Meunier, F. J., and Zylbelberg, L., 1987. On the origin of ganoine: Histological and ultrastructural data on the experimental regeneration of the scales of *Calamoichthys calabricus* (Osteichthyes, Brachyopterygii, Polypteridae), *The American Journal of Anatomy* **180:**391–402.

Slavkin, H. C., and Diekwisch, T., 1996, Evolution in tooth developmental biology: Of morphology and molecules, *The Anatomical Record* **245:**131–150.

Smith, M. M., 1995, Heterochrony in the evolution of enamel in vertebrates, in: *Evolutionary Change and Heterochrony* (K. J. McNamara, Ed.), pp. 125–150. John Wiley and Sons, London.

Smith, M. M., and Hall, B. K., 1990, Development and evolutionary origins of vertebrate skeletogenic and odontogenic tissues, *Biological Reviews* **65:**277–373.

Smith, M. M., and Hall, B. K., 1993, A developmental model for evolution of the vertebrate exoskeleton and teeth. The role of cranial and trunk neural crest, *Evolutionary Biology* **27:**387–448.

Smith, M. M., Sansom, I. J., and Smith, M. P., 1995, Diversity of the dermal skeleton in Ordovician to Silurian vertebrate taxa from North America: Histology, skeletogenesis and relationships, *Geobios, M.S.* **19:**65–70.

Smith, M. M., Sansom, I. J., and Smith, M. P., 1996, "Teeth" before armour: The earliest vertebrate mineralized issues, *Modern Geology* **20:**303–319.

Smith, M. P., Sansom, I. J., and Repetski, J. E., 1996, Histology of the first fish, *Nature* **380:**702–704.

Soler, C., and Carpenter, G., 1994, The epidermal growth factor (EGF) family, in: *Guidebook to Cytokines and Their Receptors* (N. A. Nicora, Ed.), pp. 194–196, Oxford University Press, Oxford.

Stock, D. W., and Whitt, G. S., 1992, Evidence from 18S ribosomal RNA sequences that lampreys and hagfishes form a natural group, *Science* **257:**787–789.

Stricker, S. A., 1982, Stylet formation in nemerteans, *Biological Bulletin* **162:**387–403.

Sweet, W. C., 1988, *The Conodonts: Morphology, Taxonomy, Paleoecology and Evolutionary History of a Long-extinct Animal Phylum*, 211 pp., Oxford University Press, New York.

Szaniawski. H., 1971, New species of Upper Cambrian conodonts from Poland, *Acta Palaeontologica Polonica* **16:**401–413.

Szaniawski, H., 1982, Chaetognath grasping spines recognized among Cambrian protoconodonts, *Journal of Paleontology* **56:**806–810.

Szaniawski, H., and Bengtson, S., 1993, Origin of euconodont elements, *Journal of Paleontology* **67:**640–654.

Thesleff, I., and Nieminen, P., 1996. Tooth morphogenesis and cell differentiation, *Current Opinion in Cell Biology* **8:**844–850.

Turbeville, J. M., and Ruppert, E. E., 1985, Comparative ultrastructure and the evolution of nemertines, *American Zoologist* **25:**127–134.

Turbeville, J. M., Field, K. G., and Raff, R. A., 1992, Phylogenetic position of phylum Nemertini, inferred from 18S rRNA sequences: Molecular data as a test of morphological character homology, *Molecular Biology and Evolution* **8:**669–686.

Turbeville, J. M., Schulz, J. R., and Raff, R. A., 1994, Deuterostome phylogeny and the sister group of the chordates: Evidence from molecules and morphology, *Molecular Biology and Evolution* **11:**638–645.

Urbanek, A., 1986, The enigma of graptolite ancestry: Lesson from a phylogenetic debate, in: *Problematic Fossil Taxa* (A. Hoffman and M. H. Nitecki, Eds.), pp. 184–226, Oxford University Press, New York.

Vainikka, S., Mustonen, T., and Alitalo, K., 1994, Fibroblast Growth Factors (FGFs), in: *Guidebook to Cytokines and Their Receptors* (N. A. Nicora, Ed.), pp. 215–217, Oxford University Press, Oxford.

von Bitter, P. H., and Norby, R. D., 1994, Fossil epithelial cell imprints as indicators of conodont biology, *Lethaia* **27:**193–198.

Wada, H., and Satoh, N., 1994, Details of evolutionary history from invertebrates to vertebrates, as deduced from sequences of 18S rRNA, *Proceedings of the National Academy of Sciences, U.S.A.* **91:**1801–1804.

Wade, M., 1972, *Dickinsonia*: Polychaete worms from the late Precambrian Ediacara fauna, South Australia, *Memoirs of the Queensland Museum* **16:**171–190.

Whittington, H. B., 1985, *The Burgess Shale*, Yale University Press, New Haven.

Wicht, H., and Northcutt, R. G., 1992, The forebrain of the Pacific hagfish: A cladistic reconstruction of the ancestral craniate forebrain, *Brain, Behaviour and Evolution* **40:**25–64.

Williams, A., Cusack, M., and MacKay, S., 1994, Collagenous chitinophosphatic shell of the brachiopod *Lingula*, *Philosophical Transactions of the Royal Society London B* **346:**223–266.

Williams, N. A., and Holland, P. W. H., 1996, Old head on young shoulders, *Nature* **383:**490.

Willmer, E. N., 1974, Nemertines as possible ancestors of the vertebrates, *Biological Reviews of the Cambridge Philosophical Society* **49:**321–364.

Willmer, E. N., 1975, The possible contribution of nemertines to the problem of the phylogeny of the protochordates, *Symposia of the Zoological Society of London* **36:**319–345.

Willmer, P., 1990, *Invertebrate Relationships: Patterns in Animal Evolution*, 400 pp., Cambridge University Press, Cambridge.

Winnepenninckx, B., Backeljau, T., and De Wachter, R., 1995, Phylogeny of protostome worms derived from 18S rRNA sequences, *Molecular Biology and Evolution* **12:**41–649.

Wright, K. A., 1991, Nematoda, in: *Microscopic Anatomy of Invertebrates, Volume 4, Aschelminthes* (F. W. Harrison and E. E. Ruppert, Eds.), pp. 119–195, Wiley-Liss, New York.

Yalden, D. W., 1985, Feeding mechanisms as evidence for cyclostome monophyly, *Zoological Journal of the Linnean Society* **84:**291–300.

Zhang, S., Aldridge, R. J., and Donoghue, P. C. J., 1997, An Early Triassic conodont with periodic growth? *Journal of Micropalaeontology* **16:**65–72.

Zhuravlev, A. V., 1994, Polygonal ornament in conodonts, *Lethaia* **26:**287–288.

4

Evolutionarily Stable Configurations: Functional Integration and the Evolution of Phenotypic Stability

GÜNTER P. WAGNER* and KURT SCHWENK*

Key Words: phenotypic evolution; natural selection; functional integration; stasis; constraint; Reptilia; Squamata; feeding; chemoreception; tongue; functional morphology

INTRODUCTION

Phenotypic evolution has been studied since Darwin established the fact of evolution. In contrast, molecular evolution has been a subject of study since the mid-1960s. Nevertheless, our understanding of the mechanisms of phenotypic evolution is far less developed than our knowledge of

*authorship equally shared

GÜNTER P. WAGNER • Department of Ecology and Evolutionary Biology and Center for Computational Ecology, Yale University, New Haven, Connecticut 06520-8106. KURT SCHWENK • Department of Ecology and Evolutionary Biology, University of Connecticut, Storrs, Connecticut 06269-3043.

Evolutionary Biology, Volume 31, edited by Max K. Hecht *et al.* Kluwer Academic / Plenum Publishers, New York, 2000.

molecular evolution. This fact is often attributed to the greater "complexity" of phenotypic characters, although it is not always clear what complexity means. More specifically, there are two features of phenotypic evolution that make molecular and phenotypic evolution quite distinct problems. First, molecular evolution is a continuing process, often occurring over long periods of time at a nearly constant rate, even if there are variations in rate among lineages. In contrast, phenotypic evolution is perceived as a highly irregular process with long periods of stasis interrupted by short bursts of change (Gould and Eldredge, 1977; Kimura, 1983). Second, most phenotypic characters comprise many levels of organization from the molecular to the behavioral and the population level, and the rate of change is nonuniform across these levels of organization. Some attributes of the phenotype, such as color and size, vary widely and evolve rapidly whereas other aspects of the phenotype, such as mode of food acquisition, are remarkably stable. Furthermore, even the conservative elements of the phenotype are not immutable because they have evolved in ancestral lineages and may become variable in a descendant lineage. Molecular evolution, on the other hand, pertains to evolutionary change on only one level of organization.

The fact of phenotypic stability has long been recognized and a number of mechanisms have been proposed to explain it, including developmental constraint, generative entrenchment, internal selection, functional integration, and burden, among others. In this chapter we formulate a conceptual model to explain some of the patterns of phenotypic evolution. We do not, however, propose a new mechanism for character conservation but suggest a systematic framework for the study of functional integration (e.g., Bock and von Wahlert, 1965; Dullemeijer, 1989; Wake *et al.*, 1983), internal selection (e.g., Whyte, 1965; Arthur, 1997) and related mechanisms (e.g., Simpson, 1953; Riedl, 1978; Wimsatt and Shank, 1988; Hall, 1996). We also do not review the considerable literature on evolutionary constraint and other "internal" factors of evolution, but concentrate specifically on the mechanistic issues underlying the phenomenon of phenotypic stability due to functional integration (a discussion of related concepts is given in a later section). We suggest that characters can be elements of tightly integrated clusters that we call "Evolutionarily Stable Configurations" (ESCs). ESCs are character complexes that interact in their contribution to fitness in a way that leads to evolutionary self-stabilization. Current knowledge on the evolution of lingual feeding in iguanian lizards is reviewed to exemplify the intuitive ideas underlying the concept and to propose a list of postulates characterizing the expected properties of an ESC. We then use these postulates to predict the evolutionary scenarios that may lead to the origin and disintegration of an ESC during phylogenesis. Finally we discuss how one can use character analysis and the comparative method to detect the pattern of character variation expected under the ESC model.

THE EVOLUTIONARILY STABLE CONFIGURATION: OVERVIEW

It is our thesis that some character complexes (systems) can evolve into self-stabilized, "quasi-independent" entities that resist phenotypic change, even in the face of environmentally imposed selection pressure to do so. As such, these entities remain phenotypically persistent over long periods of geological time, transcending environmental transitions, speciation, and adaptive radiations. Specifically, we argue that the nature and degree of functional integration among parts (an emergent property of the phenotype) in a given functional system (e.g., "an organ system" or "the feeding system") can influence the rate and pattern of phenotypic change in that system largely independent of any environmental selection pressures.

In our view, a highly integrated functional system is characterized by a set of functional and anatomical relationships among component parts that define the universe of adaptive variation in those parts. This limitation arises directly from the relationships themselves, and it is therefore independent of the organism's ability to build particular character states (i.e., independent of developmental constraints) and independent of external, environmental selection pressures (because the limitations are imposed by intrinsic attributes of the organism). As such, it is the internal coherence and functionality of the system, as a whole, that imposes its own "internal selection" (Whyte, 1965) on individual characters, determining which character variants are viable (Fig. 1). Internal selection is defined as differential reproductive success reflecting fitness differences among morphs that remain invariant across a range of environments (see also Arthur, 1997). By "invariant" we mean that the rank order of morphs according to fitness remains constant across environments, although absolute fitness values might vary. In addition, internal selection requires that the fitness effects of character variation depend on the other characters with which it is integrated (Kent Holsinger, pers. comm.). In short, internal selection implies rank-invariance of fitness across a range of environments and rank-dependence of fitness on the particular combination of traits posessed. Such internal selection travels with the organism wherever it goes. This is because relative fitness differences

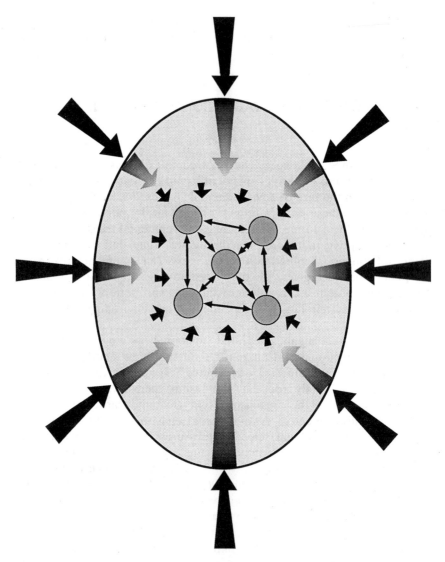

FIG. 1. Conceptual diagram illustrating the principle of "internal selection." The outer circle represents the organism and the small circles inside the "body," characters. The five characters in this figure are strongly integrated, as indicated by the many mutual interactions among them (the double-headed arrows). They thus represent an ESC. The outer, dark arrows represent selection acting on the organism from the environment ("external selection"). As the arrows continue inside the organism their intensities fade, indicating the decreasing impact of external selection on ESC characters deeply entrenched in the system. The small arrows surrounding the ESC suggest the action of internal selection resulting from the functional interdependencies of the characters. We argue, as have others, that this internal selection is the cause for the phylogenetic stability of the ESC.

are determined internally by system functionality. As su
selection exerts its influence on the phenotype of the syst
less constantly, regardless of most changes in the surrounding e
This functionally integrated, internally selected, self-stabiliz
the ESC.

During the evolution of an ESC there is a change in the dynamic between the phenotype and natural selection such that *some* phenotypic characters become less and less responsive to external selection pressures as the degree of functional integration among parts increases, while simultaneously becoming more and more sensitive to internal selection pressures coming from the relative functionality of parts within the system. As such, individual characters, or components of the system, are increasingly selected on the basis of their contribution to the relative functionality of the entire system, independent of external, environmental influences in a range of environmental conditions. It is the *system's* performance in the environmental context that is relevant, not the *character's*. Any environmental selection on the individual character is filtered through the system as a whole and is thus overwhelmingly diluted as compared to the direct selection imposed by internal system function (Fig. 1). Thus, as functional integration among parts evolves, there is a transition in the relative sensitivity of characters to different types of selection, from primarily external to primarily internal. As a result, adaptation of some characters in an increasingly integrated system occurs with decreasing reference to the external environment.

One feature of the evolution of functional systems that we do not consider explicitly here is the concept of functional trade-offs. An individual character is likely to participate in more than one function such that optimization for one function is in conflict with optimization for another. Such trade-offs can lead to functional constraint on character adaptation in addition to that stemming from the character's participation in an ESC, thus further contributing to phenotypic stability. However, the functional constraint underlying an ESC results from selection to perform a particular functional task and does not come from functional overlap. Furthermore, all constituent characters of an ESC are united by their subjugation to the same set of internal rules related to system-level function, whereas a unique set of trade-offs will potentially prevail for each individual character. Although we do not deny the possible importance of trade-offs in character evolution, our interest here is in a *particular suite* of characters. For this suite of characters, the ESC model is both necessary and sufficient to explain their stability, regardless of what other constraints might apply to individual characters. These differences clearly distinguish the ESC and functional trade-off models.

An important feature of an ESC is that it comprises a set of phenotypes that vary within certain functional parameters, rather than a single, fixed form. This is possible because of the hierarchical nature of characters and of character variation. During the evolutionary transition to an internally selected system, individual component characters become stabilized to the extent required by the functional parameters within the system. For example, one part of the system might require that two parts fit together very precisely for normal function (e.g., upper and lower incisor teeth in gnawing rodents); therefore phenotypes of these two parts must be held nearly invariant for the system to function at all (strong internal selection). However, optimal performance in a variety of environments may require modification of some other characters of the system (e.g., molar tooth form in the same herbivorous rodents), as long as character variants (states) remain within design limits imposed by system function (weaker internal selection). Whether or not variable states are adaptive for some characters within the ESC, all characters acting together must maintain the overriding function, or architectural *raison d' être* of the system (what we call the "Proper Function," following Millikan [1984]; see the text that follows) in order for the organism to survive (e.g., masticatory mode in herbivorous rodents, as in the preceding tooth examples, or tongue form in lingual-feeding lizards, as in the Case Study that follows). Therefore, we can recognize a set of characters that constitute the ESC and define or describe them at the highest level of phenotypic generality that encompasses the fundamental design principle of the system. Defined in this way, constituent characters establish an ESC "core," and it is this core that becomes "hardened" during the evolution of an ESC through the increasing action of internal selection. It is the core, therefore, that is stabilized across species and through time.

Within the functional tolerances permitted for each core character, however, variable phenotypic states of the characters can evolve. These variable states are nested within the core characters and as long as core function is maintained, their phenotypes can evolve (Appendix 2). Character states might be modified randomly through drift, or they might respond directly to external selection pressures, possibly achieving local phenotypic optima through adaptation. Because such variable states could become fixed in species or in higher taxa, the ESC can eventually encompass a set of phenotypes wherein all of which, however, maintain the fundamental, design principle (see the section titled "Character Analysis, Phylogenetic Patterns, and Comparative Methods"). Therefore, recognition and characterization of an ESC will depend on careful delineation of the hierarchical level relevant to the functional system of interest.

CASE STUDY: LINGUAL FEEDING IN IGUANIAN LIZARDS

Most ESCs will be recognized on the basis of empirical observations generated by organismal, phenotypic studies. Here we provide a case study based on the functional morphology of feeding and systematics of squamate reptiles. Specifically, we argue that functional attributes of the tongue-based (lingual) prey capture system characteristic of one clade within the group have, through functional integration and internal selection, promoted long-term phenotypic stability in the feeding system. As such, phenotypic evolution within the clade has been limited to variants consistent with the functional integrity of the prehensile tongue mechanism. Cladistic patterns of character evolution first suggested that this system was subject to some form of constraint (Schwenk, 1993). Additional analysis indicates that phenotypic stability is maintained by internal selection (functional constraint) and not developmental constraint (Schwenk, 1995a; this study, see the text that follows).

Squamate reptiles (Squamata) include lizards, snakes, and amphisbaenians. Their sister group, the Rhynchocephalia, is represented by two species of *Sphenodon* (tuatara) inhabiting coastal islands of New Zealand. Together these constitute the Lepidosauria, whose sister clade is the Archosauria, comprising crocodilians, birds, dinosaurs, and a variety of other extinct, diapsid forms (Gauthier *et al.*, 1988) [recently, however, Rieppel and deBraga (1997) and deBraga & Rieppel (1997) have argued that turtles are the sister group of Lepidosauria, but this is presently in contention (e.g., Lee, 1997; M. Lee pers. comm.). Within squamates there are two basal lineages, the Iguania and Scleroglossa (Fig. 2; Estes *et al.*, 1988). Scleroglossans are speciose and diverse, comprising several suprafamilial taxa; we do not need to consider this diversity further. Iguania includes three families: Iguanidae, Agamidae, and Chamaeleonidae (Estes *et al.*, 1988; Schwenk, 1994b; Macey *et al.*, 1997).

Iguania and Scleroglossa are distinguished by, among other things, the manner in which they procure prey/food and bring it into the mouth: iguanians use the tongue as a prehensile organ to capture small prey items, or other food, and draw it into the mouth (lingual feeding), whereas scleroglossans use the jaws and teeth for prehension (jaw feeding; Schwenk, 1988; Schwenk and Throckmorton, 1989; Figs. 3, 4). During lingual feeding, the jaws are parted as the lizard approaches the prey. The tongue tip is curled ventrally as it clears the jaw margins so that the dorsal surface of the tongue is steeply arched and presented anteriorly toward the food item as the tongue is protruded. When tongue–prey contact is made the

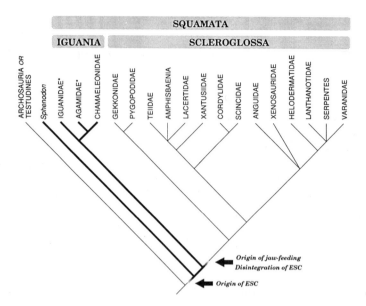

FIG. 2. Phylogenetic hypothesis underlying the case study of the evolution of lingual feeding. We propose that the lingual feeding mode found in *Sphenodon* and in the Iguania is an example of an ESC. Bold lines indicate lineages that possess the putative ESC characters. Note that the taxa exhibiting the ESC are paraphyletic because the ancestors of one descendant group, the Scleroglossa, "escaped" from the ESC. This shows that the ESC is not immutable, however escape is rare, having occurred only once in this example. Phylogenetic hypothesis is based on Gauthier *et al.* (1988); Estes *et al.* (1988); Schwenk (1988); Macey *et al.* (1997); and Rieppel and deBraga (1997).

---▶

FIG. 3. Case study phenotypes. Left column = lateral snapshots of lizards during feeding or chemoreception showing position and form of the tongue (not to scale). Middle column = kinematic plots of tongue and jaw movements during the featured behavior (not to scale). The x-axis represents time and the y-axis represents distance. Upper curve shows gape angle during prey prehension or a tongue-flick. A downward curve indicates mouth opening, an upward curve, mouth closing. Lower curve shows distance of extraoral tongue protrusion (feeding only). Right column = schematic representations of the tongue for that species in transverse section highlighting papillary form. Distribution of mucous epithelium is shown in black. Dark circle at center indicates lingual process. ESC phenotypes are shown in A–C. **A**, *Dipsosaurus dorsalis* (Iguania, Iguanidae), lingual prehension. Note curled tongue-tip, arched form of the tongue, and presentation of the dorsal surface toward the prey. **B**, *Phrynocephalus helioscopus* (Iguania, Agamidae), lingual prehension. **C**, *Chamaeleo zeylanicus* (Iguania, Chamaeleonidae), lingual prehension, at kinematic stage homologous to A and B. **D**, *Lacerta viridis* (Scleroglossa, Scincomorpha, Lacertidae), jaw prehension. Note that the tongue is retracted into the floor of the mouth. **E**, *D. dorsalis*, chemosensory tongue-flick. Note that the tongue-tip is pointed anteriorly. In a tongue-flick bout there is only a single protrusion with little bending. Compare to F. **F**, *Varanus exanthematicus* (Scleroglossa, Anguimorpha, Varanidae), chemosensory tongue-flick. Figure shows initial tongue position and maximum extension during one oscillation of a flick bout. Multiple oscillations are characteristic of many Scleroglossan tongue flicks. Note the narrow tongue, the extreme protrusion distance, and the deeply forked tip, all optimizations of vomeronasal chemoreception found only in scleroglossans. A, B based on Schwenk (unpublished data); Schwenk and Throckmorton (1989). C based on Schwenk (unpublished data); Schwenk and Bell (1988); Bell (1990). D based on Urbani and Bels (1995). E, F based on Schwenk (unpublished data).

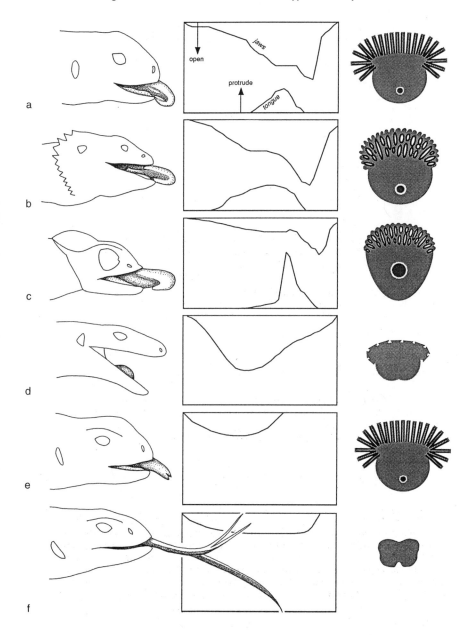

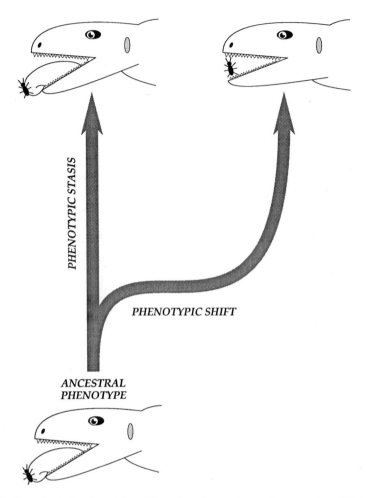

FIG. 4. Hypothesis for the origin of the scleroglossan type of prey capture. We believe that the most likely mode of escaping the lingual feeding ESC was a shift in the functional context of feeding, possibly a temporary scarcity of small prey items. According to the ESC model, such a shift would undermine the condition for its stability, namely that selection acts to optimize the proper function of the ESC. The proper function of the lingual feeding ESC is the prehension of small food items with the tongue. A lack of such prey items would create an environment outside the range tolerated by the ESC and therefore would open the door to its disintegration.

tongue is rapidly retracted into the mouth with adherent prey while the jaws open wide for clearance of the food item and then snap shut. In chamaeleonids, this kinematic sequence has been modified by the addition of a ballistic, projectile phase at the end of the tongue protrusion stage, allowing the capture of prey items as much as one and a half body lengths away (Bramble and Wake, 1985; Schwenk and Bell, 1988). In all iguanian species, the tongue comprises an effective adhesive mechanism (Tables I, II, III; Fig. 3) that is capable, in some cases, of retracting the lizard's own body weight (Schwenk, unpublished data). A single feeding event involves tightly coordinated movements of the body, head, jaws, tongue, and hyobranchial apparatus (the skeleton supporting the throat and tongue). Movement of the tongue, alone, includes whole tongue movements linked to the hyobranchium, as well as complex shape changes generated by intrinsic lingual musculature. Lingual shape changes are patterned differentially along the length of the tongue. All movements presumably reflect complex motor nerve patterning produced by putative central nervous system pattern generators (Hiiemae and Crompton, 1985). Relative kinematic patterns of all parts are highly conservative among species (Fig. 3; Bels *et al.*, 1994; Bramble and Wake, 1985; Schwenk and Throckmorton, 1989; additional citations in Tables I, II, III), reflecting the tight integration among parts necessary for a successful lingual feeding sequence, no matter what the diet, phylogenetic, or ecological background for a particular species.

The iguanian lingual feeding system is nearly identical to that of the squamate sister taxon, *Sphenodon* (Gorniak *et al.*, 1982; Schwenk, in press) and outgroup analysis shows that lingual prehension of food is primitive for Squamata (Schwenk, 1988; Schwenk and Throckmorton, 1989). As such, lingual prehension represents the retention of an ancestral feeding pattern and scleroglossan jaw prehension is an evolutionary novelty (Figs. 2, 4). Nonetheless, there is no clear dietary or other ecological variable, potentially affecting the feeding system, that uniquely characterizes either group. Within both Iguania and Scleroglossa there are dietary specialists and generalists (Greene, 1982), herbivores (including frugivores and folivores), insectivores, myrmecophages, and carnivores (e.g., Greene, 1982; Greer, 1989; Pianka, 1986; Pough, 1973). Both taxa range from temperate desert to tropical rain forest and both include fossorial (and hole-dwelling), terrestrial, saxicolous, and arboreal (broad and narrow perch) species (e.g., Bauer, 1992; Greer, 1989). Thus, origin of a novel, jaw prehension feeding system in Scleroglossa is not apparently associated with a dietary or other ecological shift or specialization (but see "Escape" scenario that follows). Despite dietary and ecological variation within Iguania, the lingual feeding phenotype (morphology, function, and behavior) remains fundamentally similar.

TABLE I. ESC Characters in Iguania and Corresponding Traits in Scleroglossa[a]

| | Iguania | | | Scleroglossa | |
Character	ESC Diagnostic Character Description	Sub-ESC Variable Character States	Putative Functional Significance	Corresponding Scleroglossan Character States	Putative Functional Significance
1. foretongue papillae	high profile	a. filamentous b. reticular	1. adhesion—frictional interlocking 2. energy absorption at prey impact 3. conversion of sheer into tensile stresses during retraction	low profile: a. loss (smooth) b. imbricate scale c. plicate d. peglike	1. decreased surface area, friction-reduction for TF 2. fluid-mechanics of molecular sampling 3. mechanics of delivery to vomeronasal organs
2. foretongue shape	high diameter: (blunt, rounded, and deep)	none	1. maximize surface area for prey contact 2. accommodate well-developed lingual musculature	low diameter: a. spatulate b. tapered c. narrow, parallel-sided d. combinations of a, b, c	decrease in diameter to increase effectiveness of hydrostatic elongation mechanism for TF (also spectacle-wiping in Gekkota and in Xantusiidae)
3. tongue tip	notched	none	unknown	variable: a. notched b. cleft c. forked d. deeply forked	1. increase turbulence during TF to enhance chemical sampling 2. tropotaxis for scent trail following during TF

4. foretongue glands	highly glandular	a. papillary surfaces b. crypts penetrate into submucosal musculature	adhesion—mucus as glue	absent	friction reduction for TF
5. foretongue surface epithelium	unkeratinized, often rugous	a. stratified squamous b. arborate	adhesion—frictional interlocking	keratinized, stratified squamous, never rugous	friction reduction for TF
6. lingual process—form	robust	a. tapered b. parallel-sided	1. tongue support 2. whole-tongue movement 3. aim tongue during prehension 4. tongue protrusion/projection	variable, mostly reduced: a. robust b. slender c. reduced d. lost	1. tongue support 2. whole-tongue movement
7. lingual process—length	long: at least 50% resting length of tongue	50–90% resting tongue length	as noted previously	short: 0–50% resting tongue length (or not in tongue)	as noted previously
8. lingual process—position	within corpus of tongue surrounded by verticalis musculature	none	as noted previously	variable: a. within corpus of tongue surrounded by verticalis musculature b. along ventral midline of tongue c. in floor of mouth ventral to tongue d. absent	as noted previously

(continued)

TABLE I. (Continued)

	Iguania			Scleroglossa	
Character	ESC Diagnostic Character Description	Sub-ESC Variable Character States	Putative Functional Significance	Corresponding Scleroglossan Character States	Putative Functional Significance
9. gape kinematics—ingestion	distinct slow-open I, SO II, fast-open, fast-close and slow-close/ power stroke phases	none	reflects precise correlation of jaw, tongue, and hyobranchial movements during prehension sequence	variable: a. cycles evident b. usually lost, modified into smooth curve	jaw movements no longer precisely correlated with tongue and hyobranchial movements during prehension; jaws open and close on prey
10. tongue kinematics—ingestion	max. tongue protrusion at end of SO II with tongue–prey contact	none	tongue protruded during SO; prey contact initiates FO and tongue retraction	tongue retracted (minor exceptions)	tongue out of way for jaw prehension; in position for intraoral manipulation and transport of prey
11. tongue form—ingestion	papillary surface protracted around end of lingual process; dorsal surface arched, tongue tip tucked ventrally	none	dorsal papillary surface presented toward prey for tongue–prey contact and adhesion on anterior third of tongue	tongue retracted, flat	as noted previously

12. tongue–hyobranchium coupling	movements coupled during ingestion, transport, and swallowing	none (uncoupled only during ballistic phase of SO II in chameleons)	most anteroposterior tongue movement is whole-tongue movement caused by hyobranchial movement	variable: a. coupled b. uncoupled during one or all feeding stages	1. ingestion does not require tongue protraction 2. hyolingual transport and swallowing replaced by specialized mechanisms such as inertial feeding
13. mesokinesis (upper jaw movement)—ingestion	absent	none	upper and lower jaws stabilized for lingual prehension	variable: a. present (most spp.) b. absent (probably in a few spp.)	jaw prehension

[a]Sources: primarily Schwenk (1988, 1994b, unpublished data); Schwenk and Bell (1988); Schwenk and Throckmorton (1989); Schwenk, Jenkins, and Sheen (unpublished data). Also: Bell (1990); Bels and Goose (1989); Bels et al. (1994); Delheusy and Bels (1992); Delheusy et al. (1994); Frazzetta (1986); Goose and Bels (1992); Herrel et al. (1995); Kraklau (1991); Smith (1984, 1986, 1988); Smith and Mackay (1990); Throckmorton (1980); Urbani and Bels (1995).

TABLE II.   Functional–Morphological Conflict in the Foretongue of
Iguanian Lizards[a]

Feature	During Lingual Prehension	During Tongue Flicking
1. protrusion mechanism	hyobranchial protraction (pushes tongue from behind) genioglossus contraction (pulls tongue from sides)	hydrostatic deformation (reduction in foretongue diameter caused by intrinsic musculature causes compensatory increase in length)
2. tongue tip form	blunted, arched, tip curled ventrally, protruded dorsum first	pointed, protruded tip first
3. relationship to hyobranchial movement	coupled	uncoupled
4. relationship to gape	wide gape to clear arched tongue and retracted prey item	narrow gape sufficient to clear reduced-diameter foretongue
5. status of high-profile papillae	positive (part of adhesive mechanism)	only negative (owing to increased friction and necessity of increasing gape to clear)
6. status of mucous glands	positive effect (part of adhesive mechanism)	negative effect (owing to increased friction during rapid TF; may interfere with fluid flow during delivery phase)
7. status of notched tip	negative effect (tucked ventrally out of way during feeding; would interfere with adhesion)	positive effect (putatively associated with chemical sampling mechanism)
8. status of dorsiventral movement during protrusion	negative effect (tongue must be held rigidly toward prey for effective targeting)	positive effect (dorsiventral oscillation putatively enhances chemical sampling)

[a] Sources: Schwenk (1993, 1996, unpublished data); Schwenk and Bell (1988); Schwenk and Throckmorton (1989); Schwenk, Jenkins, and Sheen (unpublished data); Smith (1984, 1986).

Closely related taxa have often diverged radically in diet and/or habitat without any concomitant change in the lingual feeding system (e.g., *Liolaemus monticola*, a rock-dwelling, temperate ant-specialist [Jaksic *et al.*, 1979] vs. *L. magellanicus*, a high-latitude, open grassland herbivore [Jaksic and Schwenk, 1983]), and even distantly related iguanian taxa, no matter what their diet or ecological circumstances, retain similar lingual feeding systems.

TABLE III. Foretongue Optimization for Feeding versus Flicking[a]

Lingual Prehension		Tongue Flicking	
Optimized Feature	Taxon in which Observed	Optimized Feature	Taxon in which Observed
1. *Whole-tongue protrusion mechanism* a. long, robust lingual process b. tight anatomical coupling between tongue and hyobranchium c. well-developed genioglossus muscles that insert well forward on tongue	*Sphenodon* Iguania	1. *Hydrostatic protrusion mechanism* a. narrow diameter b. tongue and hyobranchium anatomically decoupled c. well-developed circular fibers d. discrete hyoglossus bundles e. genioglossus muscle that inserts posteriorly on tongue	Scleroglossa
2. *Tongue-curling mechanism* a. genioglossus fibers run anteriorly into tip (intrinsic muscles in chameleons)	*Sphenodon* Iguania	2. *Tongue-pointing mechanism* a. hydrostatic stiffening and elongation b. posterior insertion of genioglossus fibers with no fibers running into tip	Scleroglossa
3. *Adhesive mechanism* a. long papillae to absorb impact energy; for frictional interlocking; for translation of shear into tensile stress; to form surface to prey, increase 3-D surface area of contact b. broad surface for large contact area c. surface rugosities for frictional interlocking d. copious mucus for sticking	*Sphenodon* Iguania	3. *Friction reduction* a. low-profile surface b. smooth, lack of surface rugosities c. lack of sticky mucus d. keratinized e. low surface area	Scleroglossa
4. *Targeting and stabilizing mechanism* a. long, robust lingual process in corpus of tongue b. coupling of tongue to hyobranchium c. tongue protruded on lingual process	*Sphenodon* Iguania	4. *Oscillation mechanism* a. tongue and hyobranchium decoupled b. hydrostatic stiffening c. well-developed dorsal longitudinal fibers	Scleroglossa
5. *Tongue-tip form* a. blunt, rounded b. nonbifurcate	*Sphenodon* Iguania (tip only slightly notched)	5. *Tongue-tip form* a. tapered, pointed b. forked	Scleroglossa

[a] Sources: Schwenk (1986, 1988, 1993, 1994c, 1996, unpublished data); Schwenk and Bell (1988); Schwenk and Throckmorton (1989).

Persistence of the lingual feeding phenotype throughout the adaptive radiation of Iguania cannot be accounted for by environmental stability or by commonalty of any discernible suite of external selection pressures. Despite long-term phenotypic stability in the lingual feeding character complex, there is, at the same time, compelling evidence that selection has existed historically in Iguania, and continues to exist, for the abandonment of tongue use in food prehension and therefore, disintegration of the lingual feeding system. This evidence derives from common use of the tongue by all squamates in a separate function, vomeronasal chemoreception.

Vomeronasal chemoreception is a fundamentally important sensory system for all squamates and it underlies much of their behavioral ecology (e.g., Burghardt, 1970; Cooper, 1996; Halpern, 1992; Mason, 1992; Schwenk, 1995b). During vomeronasal function, tongue protrusion ("tongue flicking"; Fig. 3) is used to capture molecules and to carry them into the mouth for delivery to paired chemoreceptors (the vomeronasal organs) above the palate (Halpern, 1992; Schwenk, 1995b, 1996). Iguanians, as well as scleroglossans, use tongue flicking and vomeronasal chemoreception; however, it is relatively more poorly developed. For decades iguanians were regarded as a "visual" clade in contrast to the higly chemosensory scleroglossans. Although this dichotomous view was exaggerated, it reflects the greater development of the vomeronasal system in the Scleroglossan clade (reviewed by Schwenk, 1994a). In any case, it is only in scleroglossans that the system has become extremely specialized and diverse. In some scleroglossan taxa, especially snakes, the tongue–vomeronasal system reaches its greatest functional complexity (Halpern, 1992; Schwenk, 1994c, 1995b). Specialization of the vomeronasal system in scleroglossans is correlated, in particular, with numerous modifications of the foretongue that optimize tongue-flick behavior, including narrowing and thinning of the foretongue, reduction and loss of surface papillae, loss of lingual glands, and forking of the tip (Schwenk, 1988, 1993, 1994c, unpublished data; Tables I, III). These modifications eliminate the mechanical ability of the tongue to act as a prehensile organ during feeding because they disrupt the critically important adhesive mechanism. Tongue flicking requires a different tongue conformation during protrusion and presumably different central patterning and motor control (Fig. 3). Consequently, foretongue optimization for lingual feeding is in *direct conflict* with optimization for chemoreception (Tables II, III). The ability of iguanians and their lingual feeding ancestors to enhance vomeronasal chemosensory function by adaptively modifying the foretongue has been prevented by the functional/biomechanical demands of the integrated lingual feeding system on tongue form (Schwenk, 1993, 1995a). Iguanians lack functional specializations of the vomeronasal system characteristic of scleroglossans (Halpern, 1992; Schwenk, 1993, 1994c, 1995b). However, once a new mechanism of prey prehension evolved (use of the jaws; see the following text) in ancestral scleroglossans, the tongue was free to respond to preexisting selection pressure for chemoreception and the foretongue could be modified and optimized for tongue flicking and for vomeronasal function (Schwenk, 1993, 1995a).

We postulate that strong selection exists in iguanians, as in scleroglossans, for phenotypic modifications that enhance vomeronasal function, but that mutations leading to such modifications are selected against owing to the fitness cost associated with impaired functionality of the lingual feeding system (Table II). Selection against chemosensory adaptation in iguanians must arise internally from feeding system function and integration among parts, even when potentially strong external (environmental) selection exists for modification of unit characters (such as tongue surface characters; Table I) within the system. Internal selection has circumscribed the universe of allowable variation in any given lingual character, restricting it to forms that enhance, or are neutral to, the existing function of the character complex, in this case lingual food capture.

We propose that the iguanian lingual feeding system constitutes an ESC for the following reasons:

1. It includes a set of phenotypes that perform a single proper function (Millikan, 1984; Appendix 1) that can be characterized as "the prehension and delivery of food into the mouth."

2. All species evincing the system have a common ancestor and therefore the lingual feeding character states are part of a mutationally connected set.

3. It comprises numerous anatomical parts, or characters, that must operate together in a carefully correlated kinematic sequence for the proper function to occur (Table I). The lingual feeding system is highly integrated and this integration imposes its own internal selection on character change. Transitions from one character state to another (e.g., reduction in papillary length, reduction in lingual mucous glands, or hyobranchial reduction), even those that might enhance some other function, such as vomeronasal chemoreception for finding food or mates, will have negative fitness consequences if they diminish feeding performance.

4. Fundamentals of the morphology, kinematics, and behavior of lingual feeding are stable across the entire clade, independent of environment. In this example, the ESC-associated environments are diverse, though finite, being all of those that provide food and other resources required by lizards. Given that this system represents retention of the ancestral feeding pattern,

we infer that it has remained stable throughout the adaptive radiation of Iguania, at least since the early Jurassic, circa 190 mybp (Estes, 1983), and probably far longer because the presence of a nearly identical system in the extant sister taxon, *Sphenodon*, suggests an even more ancient origin (at least early Triassic; Carroll, 1988) of the lingual feeding mechanism.

5. There is evidence for historical and ongoing selection for character modifications that would disrupt the lingual feeding phenotype, but that has not, except for the single case of the origin of Scleroglossa (see "Escaping From an ESC").

Chameleons

The chamaeleonids are unusual among iguanians in having evolved a derived, projectile tongue apparatus. Here we discuss this case as example of a highly specialized form evolving within the putative lingual feeding ESC of iguanians. It is important to recall that an ESC constitutes a set of phenotypes that are stable in some diagnostic, or fundamental suite of characters (the core), but might vary in subordinate characters or character states. As such, for a given proper function performed in a variety of habitats under diverse ecological conditions, the optimal performance of the proper function might require a number of special modifications. This implies that even if we expect conservation of the core ESC characters, the evolutionary pattern of the ESC subordinate characters cannot be predicted. Subordinate character variation may be completely unconstrained, leading to frequent homoplasy in character states, or it may be restricted for a variety of reasons (see Appendix 2).

There are at least four reasons for a strong phylogenetic pattern in variation of ESC subordinate characters: novelties, sub-ESCs, functional trade-offs, and canalization. Novelties may induce a strong phylogenetic signal on variation in ESC subordinate characters if the probability of invention is low. For example, agamid lizards have a unique form of the anterior lingual papillae and glands such that the papillae are anastomosed along their lengths so that the covering glandular epithelium is formed into convoluted crypts that, in some places, penetrate deeply into the corpus of the tongue musculature (Schwenk, 1988). The functional basis of this novel configuration, if any, is unknown, but it seems to have arisen only once in the ancestor of Agamidae and is retained, albeit slightly modified, in the Chamaeleonidae. Another possibility is that the ESC-subordinated characters themselves form a "sub-ESC" in a part of the ESC taxon. We do not know of a likely example but mention it as a possibility. Furthermore, one has to keep in mind that the ESC subordinate characters may have functional interactions with other systems (trade-offs). If there is an adaptive trend in these characters it may be that they influence the evolution of the ESC-subordinate characters. For example, although we argue that the derived chamaeleonid condition retains membership within the ESC as a lingual, prehensile system (see further discussion in Appendix 2), its evolution may have been influenced by such trade-offs. The chamaeleonid system is apparently optimized for an extreme ambush predation strategy coupled with the need to remain stationary while capturing prey owing to a physiological incapacity for rapid locomotory movement (Abu-ghalyun, et al., 1988); the need to maintain crypticity; limitations of the structural habitat, especially occupation of narrow, three dimensionally complex perches; and/or the need for the body to be properly oriented in order to resist the inertial reaction forces of tongue projection and the unpredictable forces of pendulatory tongue movements with attached prey during retraction, all of which would tend to destabilize the chameleon on its narrow perch. In generalized, nonprojectile lingual feeders that eat comparable food types (Iguanidae and Agamidae), prey are ambushed with a short, rapid sprint. Thus, in chameleons, specialization of the prehensile tongue mechanism is characterized by many autapomorphic character state modifications in the hyolingual system that are associated with adaptation to a specialized habitat. This association leads to a conservation of subordinate ESC character states among the chameleons. The last possibility for the conservation of subordinate ESC phenotypes is that characters may become developmentally canalized due to extended periods of stabilizing selection (Wagner et al., 1997). Internal selection on ESC characters could provide such long-term stabilizing selection on individual characters. Again, we are not aware of an example for this possibility but think that it is highly plausible. We explore all these possibilities in detail in the section titled "Character Analysis, Phylogenetic Patterns and Comparative Methods."

THE ESC: CHARACTERIZATION AND POSTULATES

An ESC is a configuration of morphological, physiological, and behavioral characters that are adaptive in a range of accessible habitats. The scale of an ESC is variable, potentially ranging from just a few characters to many characters (e.g., in a large, complex organ system). However, it is diagnosed by a pattern of functional integration among characters, thus it must always represent a character "complex." The character concept itself is therefore not considered here (see the volume edited by Wagner, 2000). Further, the

ESC comprises a set of states of the character complex, each of which is adaptive within an associated habitat, but that nonetheless maintain a certain design principle. The set of character states included within an ESC is mutationally connected and adaptively closed; that is, the ESC represents a natural, historical entity. In sum, an ESC is best seen as reflecting the three-way interaction among the morphological–physiological substrates of function, the behavioral matrix in which the morphological–physiological characters are embedded, and the range of habitats occupied by species exhibiting the ESC. Given that the idea of an ESC is derived from a few empirical examples, we do not feel comfortable providing a formal definition. We believe that a formal definition given too early in the development of a concept can stifle its further development. We prefer, instead, to propose a brief characterization and a series of descriptive postulates that follow:

An ESC is a set of adaptations (occurring in different species) that performs a given proper function (*sensu* Millikan, 1984; Appendix 1) and that is closed under adaptation.

In short, an ESC is a closed set of states of a character complex devoted to a particular proper function. This set of states is closed under the action of natural selection in a range of environments. This statement requires a few comments. The term "closed set" is borrowed from mathematics, where it is used extensively in algebra. Consider, for example, the rule of algebraic addition, such as $4 + 2 = 6$. This equation says that, given any two integer numbers (e.g., 2 and 4), then application of the operation "+" obtains another integer number, in this case 6. A closed set within the set "integer numbers" is the subset of integers that yield only members of the subset under the application of "+." For instance, if one starts with two even numbers, such as 2 and 4, then the application of "+" will also yield an even number, that is, 6, a member of the same set. Hence, one can say that the set of even numbers is "closed" under the application of "+"; one never leaves the set by any application of this operation. On the other hand, the set of odd numbers is not closed under "+." Consider 3 and 5, for example. In this case, addition leads to $3 + 5 = 8$; that is, the addition of two odd numbers yields an even number and the set of odd numbers is not closed under the operation "+."

Closed sets have a similarity to what we wish to capture with the ESC concept. We have certain elements, in our case phenotypic character states, that are "transformed" by natural selection into other characters states. What we propose to call an ESC is then a set of character states that is closed under the action of natural selection, which means that natural selection only leads to elements of the same set. An ESC is something like a "trap" or a stable "orbit" into which the process of adaptation by natural selection may fall. Of course, this "adaptive orbit" cannot be stable for arbitrary selection regimes, otherwise innovation would be impossible. It can be stable for all adaptations to perform a certain proper function, such as mastication or prey capture. In the following sections we consider a list of biological assumptions, or conditions, that we think are sufficient to cause a set of character states to be a "stable adaptive orbit," that is, an ESC. The question of whether these conditions are also necessary is more difficult to assess, let alone prove. This list also provides us with a heuristic to identify the conditions under which a lineage can escape from an ESC.

Postulate 1: Mutational Connectedness

The character states in the ESC are mutationally connected. It must be possible to evolve from one character state to another without leaving the ESC. This implies that character complexes that perform the same proper function, but which are not mutationally connected, are not members of the same ESC. For example, the masticatory apparatus of both rodents and artiodactyl ungulates perform the same proper function (crushing and grinding difficult-to-digest plant matter), but these systems could not belong to the same ESC because their constituent character states are not mutationally connected. Each is adaptively isolated from the other, constituting its own closed set of locally optimized phenotypes. This means that a transition from a rodent to an artiodactyl type of mastication would include character states that are not as well-adapted for dealing with tough plant material. This is evident historically by the fact that if we follow each clade deeper into mammalian phylogeny to their common ancestor, we must pass through species whose masticatory apparatus did not share the same proper function (e.g., insectivory). As such, the rodent and artiodactyl ESCs were independently evolved and their character states are not mutationally connected; that is, one has to pass through states belonging to neither ESC in order to reach one from the other.

Postulate 2: Functional Uniqueness

The elements of the ESC are the only characters that perform a proper function in a range of associated environments. The notion of proper function is elaborated in Appendix 1. An example of a proper function is to gain access to a certain resource that is available in a set of habitats. There is only one ESC performing a given proper function within an organism. Violation of this postulate is one possible mode of escaping from an ESC (see the section titled "Escaping from an ESC").

Postulate 3: Functional Integration

Each character in the ESC is essential for performance of the proper function and each character can operate only in the context of all other constituent characters of the ESC in performing the proper function. Contributions of different parts of the ESC character complex are not additive in their contribution to performing the proper function. Epistasis in fitness is expected, which makes the function of the whole dependent on the presence of all other parts. This postulate requires special attention with respect to the origin of ESC (see the following section on the origin of an ESC).

Postulate 4: Trade-offs in Performing Other Functions

The fitness gain from improving the performance of a function other than the proper function is less than the fitness loss associated with the drop in performance of the proper function. This postulate assumes that there is an inherent conflict in optimizing more than one function with a single (common) set of characters. In other words, a mutation leading to an improvement of a function other than the proper function of the ESC will have unconditionally lower fitness in the ESC-associated environments and will be selected against. This postulate suggests a possible avenue of escape from an ESC through certain environmental changes (see "Escaping From an ESC").

Postulate 5: Functional Closure

The mutational neighborhood (i.e., outside the ESC) of the character state set is structurally disadvantaged in performing the proper function in the associated environmental conditions. A mutation or set of mutations that place the phenotype outside the ESC will be at a fitness disadvantage in comparison to ESC-resident phenotypes in the population because it will be functionally inferior in performing the proper function of the ESC. As such, within the environments associated with the ESC, adaptive evolution to optimize the proper function will remain within the ESC (the ESC is a closed set of phenotypes under adaptation). Thus, in all ESC-associated environments, all character states in the mutational neighborhood of the ESC have a lower fitness than the mutational surface of the ESC itself.

ORIGIN OF AN ESC

The origin of an ESC is paradoxical in that a conservative structure emerges from a dynamic process. How can the mechanism of mutation and selection lead to a product that is buffered from the influence of the very same mechanisms that created it? The origin of an ESC shares this problem with other conservative patterns in evolution, such as body plans (*Baupläne*) and evolutionarily fixed traits. Currently there are three models proposed to explain the evolution of conservative structures: the evolution of developmental constraints (Stearns, 1993; Wagner, 1986); the evolution of functional dependencies (burden; Riedl, 1978; Stearns, 1993); and generative entrenchment (Wimsatt and Schank, 1988). Because the stability of an ESC is assumed to be due to selection rather than developmental constraints, the first model is not relevant in this context. The second and the third, however, are both models of character conservation that rely on natural selection and are thus candidates for the origin of ESCs.

The functional burden model is based on the fact that characters can increase in evolutionary stability as organisms become more complex. For example, the axial organs of chordates (the neural tube, chorda, and somites) are variably present among basal chordates but are absolutely fixed among vertebrates. Riedl (1978) explained this increase in evolutionary stability by the higher number of characters in vertebrates that are functionally and developmentally dependent on the axial organs. Through the addition of new characters their functional importance (their burden) increases and thus they become "locked in," because their loss would have consequences for many other adaptive characters. This model may also apply to the functional integration of ESC characters. Initially the characters may be loosely integrated, but as the characters mutually adapt to one another, the performance of the whole system becomes dependent on the tight fit among all of its parts. For instance, the rodent and the ungulate herbivorous feeding systems differ, among other things, in the way the mandible is moved during mastication (primarily longitudinal in rodents and primarily lateral in ungulates). This difference is fixed, in part because of the adaptation of tooth form to the preferred direction of jaw movement: raised enamel ridges (lophs), separated by softer, lower regions of dentine, are oriented in both groups perpendicular to the direction of mandibular movement (hence, transverse in rodents, longitudinal in ungulates) in order to maximize the efficacy of grinding. Although it is possible that a more generalized tooth form (e.g., enamel-crowned, bunodont molars) might permit mandibular movement in several directions, transition to this form is not an option once tooth shape has adapted to the preferred direction of mandibular movement because there would be a necessary decrease in grinding efficiency and therefore, fitness consequences. Hence, the addition of new characters and the mutual coadaptation among them might directly lead to functional integration and to the conservative tendencies of an ESC.

The model of generative entrenchment is based on the fact that the intensity of selection for or against a variant phenotype depends on the fitness difference relative to the most fit morph in the population. This means that as deleterious mutations accumulate and the mean fitness decreases, the relative fitness loss of yet another mutation of the same absolute magnitude increases. Wimsatt and Schank (1988) have shown by computer simulation that this process leads to the preferential loss of low-impact genes or characters and the preferential retention, and ultimately the fixation, of slightly more important genes or characters. In the context of ESC evolution one can envision an ancestral condition in which the proper function was performed by a diffuse and loosely integrated system with much functional redundancy. However, as the less important characters contributing to functional performance are pruned away, the relative importance of the remaining ones increases and eventually they "crystallize" to form the core of an ESC. Clearly this model is not in conflict with the burden model, but it points to another mechanism that can also contribute to the formation of an ESC.

The common feature of the burden and the generative entrenchment models is that the conservation of characters and character states is not due to particular evolutionary events but is due to the systems-level consequences of fairly ordinary evolutionary steps. In the case of the burden model, the evolution of new adaptations causes the character on which this new adaptation is predicated to become more conservative or harder to change. In the case of the generative entrenchment model, with the accumulation of mildly deleterious mutations the adaptive importance of the remaining traits or genes increases. Hence a key feature of organismal evolution is the relationship between elementary mechanisms, like mutation, drift, and selection, and their systems-level consequences.

ESCAPING FROM AN ESC

An ESC promotes long-term phenotypic stability due to relative insensitivity to changes in environment. Nonetheless, it is apparent from the very diversity of life that there are circumstances in which it is possible for a lineage to escape from the evolutionary tyranny of an ESC, even

if these circumstances arise only rarely for a given ESC in a particular lineage.

We offer several scenarios for escape from an ESC. *The main prediction of the ESC concept is that transitions from one functional type (i.e., one
ESC) to another are **not** driven by selection for improving the given proper
function of the ESC, but by evolutionary events unrelated to the proper function.* The characterization of the ESC concept provided previously defines
the kinds of historical events that might destabilize an ESC. Among the five
postulates discussed previously, Postulates 2 and 3 are the most relevant in
this context. Postulate 2 states that the characters in the ESC are optimized
by natural selection to perform a certain proper function and that they are
the only system performing this function. Furthermore, it assumes that the
proper function contributes to fitness in a limited set of environments. Postulate 3 assumes that all of the parts contributing to the proper function are
necessary. These postulates have three key features that are the basis for
the escape scenarios discussed in the following sections:

1. The ESC is only a part of a larger system in which it contributes to
   fitness, but is not the only determinant of fitness.
2. An ESC is stable as long as there are no alternatives available to
   perform the proper function of the ESC (redundancies).
3. Performing the proper function of an ESC is adaptive only in a
   limited set of environments.

It is likely that there are other possibilities for escape from an ESC,
but these are the structural instabilities built into the very concept. Our scenarios are not mutually exclusive, but rather represent "modal" pathways
for escape. Note that, for purposes of argument, we assume our illustrative
examples to represent ESCs, but with the exception of the Case Study
example, support for this assumption is beyond the scope of this chapter.
The authors of the specific cases cited may not necessarily agree with our
characterization.

Overriding Selection Pressures and Pleiotropic Effects

The key to understanding the inherent lability of an ESC, in spite of
its conservative tendencies, is that its characters are part of an organism
that has to perform additional functions (other than the ESC proper function) to have nonzero fitness. *The stability of the ESC is only claimed with
respect to adaptations to perform its particular proper function.* However,
evolution is more than what happens in one organ system or character
complex. Consequently, all selection that acts to maintain the ESC can be

overridden by strong selection on other organismal attributes. These scenarios can roughly be classified into two classes, one in which the selection pressure acts on some global organismal property, such as developmental timing, body size, or buoyancy (*systemic influences*), and those where evolution of another, localized organ system has side effects on the character states of the ESC that may push the system beyond its limits of stability (*pleiotropic influences*).

Systemic Influences

An example of a systemic effect that may destabilize an ESC is strong selection for short developmental time, or a switch from larval to direct development. In this scenario, one possibility is paedomorphic evolution such that the repatterned, juvenilized phenotype of descendant species fails to develop the ESC character complex in the first place. The more generalized (with respect to the ESC) paedomorphic descendant is now capable of responding to external selection pressures in novel ways. Subsequent adaptive evolution can rebuild a new set of integrated systems unfettered by the original ESC phenotype. In this case, a fundamental shift in environmental demands changes the level of selection such that existing internal functionality rules are simply overridden.

Roth and Wake (1985) made a similar point regarding the evolution of feeding mechanisms in plethodontid salamanders. In their reconstruction of the evolutionary events leading to the origin of the highly specialized tongue feeding system in this group, tongue projection was made possible by historical events unrelated to feeding. For example, the evolution of direct development in the lineage Bolitoglossini was considered by Roth and Wake a necessary precondition for the evolution of highly projectile tongues.

Pleiotropic Influences

In a recent paper, Rowe (1996) provided comparative and ontogenetic data supporting the notion that the mammalian jaw joint evolved as a side effect of volume change in the telencephalon. In "reptilian" ancestors of mammals (early synapsids), the elements of the middle ear are part of the jaw joint, performing both the functions of sound transmission and jaw movement. This configuration was remarkably stable phylogenetically until the mammalian lineage escaped from this putative ESC. According to Rowe, the reason was not direct adaptation for either sensory or for jaw function, but a change in skull shape induced by the expanding neocortex. A consequence of this expansion was a spatial separation of the small, ancestral jaw joint bones (which later became the mammalian middle ear ossicles) and the larger, toothed dentary of the lower jaw, the only bone to remain in the mammalian mandible. As such, according to this interpretation, the evolutionary change in the ancestral jaw joint (the ESC) was not driven by an adaptive change in the proper function or its constituent characters (the bones making up the jaw joint) but by a functionally unrelated adaptation in another functional system, the brain and neurocranium.

Another possible example is suggested by recent work of Galis (1992) and Galis and Drucker (1996) on the origin of the cichlid fish pharyngeal jaw apparatus. According to Galis and Drucker, the ancestral state of the pharyngeal jaw is one where upper and lower pharyngeal jaws are mechanically coupled such that movement of upper and lower jaw elements cannot occur independently and bite force is dissipated. In cichlids, upper and lower pharyngeal jaws are decoupled and are capable of independent movement. Independence of upper and lower jaws allows full transmission of muscle force to the lower jaw and the exploitation of neurocranial reaction forces on the upper jaw so that cichlids achieve a powerful bite and greater flexibility. Even if the cichlid condition is mechanically superior, the putative ancestral configuration is stable in the sense that there are recent fishes that retain it (e.g., Centrarchidae). Nor is the ancestral state associated with a particular environment. Anatomically the transition requires a number of mechanical decouplings, each of which seems not to be an immediate improvement of biting function. Hence the transition to the cichlid condition is unlikely to have occurred by direct selection for improving the proper function of the pharyngeal jaws, even if in the end it led to an improvement in jaw function. Most likely the transition was initiated by a change in the shape of the skull base with its geometric and mechanical consequences (Galis, pers. comm.). Again, a major, but phylogenetically rare, transition was not due to selection for improving the proper function directly but was due to the pleiotropic and perhaps epigenetic effects of other adaptive trends functionally unrelated to the focal characters (Hall, 1992).

Redundancy in Performing the Proper Function

Functional integration promotes stability of the ESC; that is, if all parts of the system are essential for its function they tend to form a conservative complex. Functional redundancy, however, can loosen the integration of the parts and may lead to new avenues of adaptive evolution. Duplication, either of genes (Ohno, 1970; Hughes, 1994; Nowak et al., 1997) or of body parts (Riedl, 1978; Weiss, 1990; Müller and Wagner, 1991), is increasingly viewed as a significant source of phenotypic novelty. Duplicated units are free to evolve novel functions and specializations.

In the case of an ESC, duplication can come about in two ways: either the entire ESC phenotype is physically duplicated within the organism, or performance of the proper function is duplicated by another system. In the former case, duplication of a complex ESC would probably be restricted to those cases in which the ESC is expressed wholly within a serially homologous unit like a segment; addition of body segments is a well-known evolutionary phenomenon (e.g., Goodrich, 1906, 1913; Minelli and Peruffo, 1991). In the latter case, some other system or body part(s) might duplicate the proper function of the ESC if there is an appropriate shift in the environment. It would be most likely to occur if the accessory system arises initially for other purposes as a preadaptation (exaptation). For example, in the Case Study scenario, the ESC proper function was duplicated when jaw prehension of prey became adaptively favored over tongue prehension owing to a putative shift in available prey size (this is elaborated in the following text). Duplication of the proper function "permitted" fragmentation of the ESC character complex and annexation of these characters for adaptation in other systems/functions, in this case, vomeronasal chemoreception.

Similarly, the evolution of tetrapods and of aerial respiration was only possible because of duplication of respiratory function in ancestral fishes. If we accept the branchial (gill) complex as an ESC, then the evolution of lungs as accessory aerial respiratory structures duplicated its proper function (gas exchange). Accessory respiratory organs have arisen repeatedly in several fish lineages as elaborations of preexisting, well-vascularized mucosae, including the mouth epithelium of mudskippers, the suprabranchial organs of anabantoids and siluroids, and gut-breathing in the Gobitidae (Graham, 1997). As in dipnoan lungfish, accessory respiratory structures promote survival of fish during periods of hypoxia (and/or high pCO_2), or during periods of aestivation in mud. In some cases, air-breathing structures even permit sojourns of fish onto land (e.g., mudskippers and the siluroid catfish, *Clarias*). Lungs develop as outpocketings of the anterior gut (e.g., Walker and Liem, 1994) and it is likely that they evolved initially as adaptations for gut breathing in fish. A more permanent change in the environment of tetrapod ancestors (e.g., increasing frequency of anoxia) might have shifted the balance in respiratory system use so that lungs became the principal, rather than an accessory, site of gas exchange (Liem, 1988). Duplication of the gill-ESC proper function followed by a critical change in external conditions could once again have precipitated dissolution of the ESC. Once lung breathing made redundant the proper function of the gill ESC, functional integration in the branchial complex was no longer internally selected and its component parts were freed for adaptive modification into the diverse branchial arch derivatives of tetrapods (e.g., thyroid gland, hyobranchium, laryngeal cartilages, middle ear cavity and ossicles, tracheal rings).

A final example of functional redundancy might be the evolution of locomotor modes in birds. Gatesy and Dial (1996) showed that theropod dinosaurs ancestral to birds were obligate bipeds that had a single "locomotor module" comprising a functional coupling of the hind limbs and the tail, in our interpretation an ESC. Indeed, Gatesy and Dial noted the remarkable evolutionary stability of the theropod system and the failure of this system to diversify into multiple adaptive types. Early birds developed an additional locomotor module when they began to exploit the forelimbs for flight, thus introducing locomotor redundancy. Redundancy in the proper function of the ancestral locomotor ESC relaxed functional integration between tail and hind limbs such that these elements disintegrated into separate locomotor modules in birds. The increased degrees of freedom permitted by the presence of three locomotor modules in birds (forelimb, hind limb, tail) as compared to one (hind limbs plus tail) in ancestral theropods is argued by Gatesy and Dial (1996) to have permitted the great locomotor (and morphological) diversity of birds (*sensu* Vermeij, 1974; Lauder and Liem, 1989).

The common element in these scenarios is that once the proper function is duplicated by another system, internal selection on ESC characters is relaxed because there is no longer a need for them to remain optimally integrated as a single system. The characters are then able to respond to other sources of selection and integrity of the ESC is lost.

Environmental Change and Modification of Internal Optimality Rules

An ESC, as conceptualized here, is adaptive within the confines of a potentially large but nevertheless limited range of environments. Only within this limited range of environments do the functional interactions among the component characters make adaptive sense and therefore, within these environments the characters and their interactions are selectively maintained. An environmental change may remove a key provision necessary to be within "adaptive reach" of a given ESC. If this happens, a cascade of adaptive changes may be triggered, either fragmenting the ESC or potentially leading to another ESC. Details of this environment–function interaction are case-specific and therefore are best illustrated with examples: eye reduction in fossorial or cave-dwelling organisms and the evolution of jaw feeding in squamate reptiles. For the latter we refer to the Case Study outlined previously and propose a hypothetical scenario for the origin of the Scleroglossa and its escape from the ancestral, lingual-feeding ESC.

It is well known that organisms that become fossorial, caved welling, or otherwise occupy lightless adaptive zones progressively lose their eyes (Walls, 1942). Walls, for example, used the simplified eye form of snakes to argue that their ancestry included a period of fossoriality. Many cave invertebrates, fish, and salamanders are virtually eyeless. Particularly well known is the case of caecilians (Wake, 1985). Caecilians are limbless amphibians (Gymnophiona) highly adapted in all aspects of their phenotype to a fossorial existence (though some species are secondarily aquatic). Universally they exhibit eye reduction, but the nature and extent of reduction vary among species (Wake, 1985). Salient here is that caecilians uniquely possess a paired structure known as the tentacle, associated with the vomeronasal chemosensory system, used to probe their fossorial environment (e.g., Billo and Wake, 1987). The tentacle uses several structures ancestrally associated with the eye (e.g., the levator and retractor bulbi muscles of other amphibians are used by caecilians to move the tentacle; the Harderian gland that ordinarily moistens the eye is used to lubricate the tentacle; the tentacular aperture is homologous to the interpalpebral space; Wake, 1985, 1992; Billo and Wake, 1987). Indeed, in one family the eyeball itself is displaced from the orbit and extruded along with the tentacle (Wake, 1985; O'Reilly et al., 1996). Billo and Wake (1987) and Wake (1992) argued that evolution of the novel tentacular apparatus of caecilians was permitted by the regression of the eye following a shift in the ancestral habitat from terrestrial to fossorial. In our parlance, the eye can be viewed as an ESC composed of many individuated characters (see Wake, 1985 for details on mosaicism of eye character reduction) that was stable in a large number of (lighted) environments. For whatever reasons, a habitat shift occurred that placed the ESC outside this range of environments, allowing its disintegration. A new suite of external selective pressures favored elaboration of chemoreceptors adaptive in the new, fossorial environment for which the liberated components of the eye were simultaneously available. The key element here is that a particular environmental shift changed the "rules" governing stability of the ancestral ESC. It is likely that evolution of the tentacular apparatus was rapid and has been itself stabilized by a new set of internal, integrative rules. Of course, few characters function in only a single ESC and they are not necessarily completely free to evolve even if the proper function is eliminated. For example, Wake (1985) noted that all caecilian eyes appear to retain photosensitivity, possibly relating to occasional surface activity for feeding. Crandall and Hillis (1997) found that molecular evolution of the visual pigment rhodopsin remains constrained in cave-dwelling crayfish, despite the total absence of light in their environments, possibly by some additional, unknown function of the pigment (such as regulation of circadian rhythms).

The iguanian lingual-feeding ESC functions effectively over a wide range of prey mass, from ants to birds and other vertebrates. Most species take a diversity of prey (Greene, 1982), but small prey are the rule among iguanians and other lizards (e.g., Pough, 1973; Losos and Greene, 1988). Perhaps the only dietary distinction that can be made between lingual-feeding iguanians and jaw-feeding scleroglossans is that some scleroglossan lineages have evolved strategies for eating very large prey relative to body size. Snakes are a notable example, as are several lizard taxa scattered among the scleroglossan clade. Ingestion of large prey is by no means a universal trait of Scleroglossa, but as a generalization it is true that the largest prey eaten by iguanians do not approach the largest ones taken by scleroglossans in relative mass, and virtually no iguanians feed exclusively on large prey, as do some scleroglossans. One reason for this is probably the mechanical limitation of the lingual adhesive mechanism in overcoming the inertia of very heavy prey during capture and retraction into the mouth. Use of jaws and teeth for prey prehension by Scleroglossans provides greater force for grasping and lifting. Indeed, species such as snakes and some monitor lizards (*Varanus*) that eat particularly large prey items have also developed novel mechanisms for transporting and swallowing prey (Gans, 1969a; Cundall, 1995; Kley and Brainerd, 1996), functions that in other species depend on the tongue as well. Schwenk and Throckmorton (1989) noted that iguanians eating very large prey appeared to use the jaws for prehension, an apparent switch in feeding mode that was also noted by Gorniak and Colleagues (1982) for *Sphenodon*. Nonetheless, Schwenk and Throckmorton (1989) observed that, even while feeding on large food items, iguanians protrude the tongue and contact the item with the dorsal, adhesive surface. However, the distance of tongue protrusion is modulated with prey size so that for large items, tongue–prey contact occurs at the jaw margins simultaneously with jaw contact. Similarly, herbivorous species that crop free-standing vegetation with the teeth were seen to protrude the tongue little beyond the jaw margins. Schwenk and Throckmorton (1989) therefore proposed a quantitative, rather than a qualitative, shift in feeding kinematics related to prey size in these species.

How was the mechanically effective, intrinsically stable, lingual-feeding ESC circumvented in the scleroglossan lineage? Ancestral scleroglossans might have retained lingual feeding for some time after their cladogenesis. However, a key environmental shift to a set outside the ESC-associated conditions could have rapidly changed the internal functionality rules governing the lingual-feeding ESC. Specifically, an environment in which relatively small prey items were scarce or unobtainable would have required consistent use of the lingual feeding mechanism at one extreme of its kinematic continuum such that tongue–prey contact would always occur at the jaw margins with simultaneous jaw/tooth contact (as noted previously). Once this kinematic sequence became the rule rather than the exception, selection optimizing the ability of the jaws and teeth to grab prey would prevail, introducing functional redundancy that violates Postulate 2 for the stability of an ESC. Because prehension with jaws and teeth is presumably more effective for large, heavy prey (because it is not constrained by mechanical limitations of the lingual adhesive mechanism), selection would tend to optimize jaw feeding, ultimately eliminating the now redundant kinematic pattern of tongue protrusion and prey contact. Once the foretongue was no longer required for prey prehension, the set of functionality rules governing the lingual feeding mechanism, the basis for internal selection of the ESC, would be eliminated. The components of the ESC, including the tongue, would then be free to respond to other (external) selection pressures, such as vomeronasal chemoreception (Figs. 3, 4). Adaptation of the foretongue for tongue flicking, chemical sampling, and chemical delivery to the vomeronasal organs would have rapidly eliminated the adhesive mechanism required for lingual prehension (Tables I–III) and scleroglossans would have become committed to jaw prehension of prey even if subsequent environmental shifts allowed the reacquisition of small prey as part of their diet.

CHARACTER ANALYSIS, PHYLOGENETIC PATTERNS, AND COMPARATIVE METHODS

Given that an ESC is an historical entity, we expect the distribution of ESC character states on a phylogeny to show some pattern. These patterns might suggest methods for identifying ESCs and provide clues to the conditions of their evolution and disintegration.

Phylogenetic Signature of an ESC

Because both origin and escape from an ESC are relatively rare, ESC character states should be representative of a higher clade or significant parts of it. Representation of the ESC across a phylogeny should be monophyletic, or, because it is not impossible to leave the ESC, paraphyletic (as in the Case Study example, Fig. 2). Polyphyletic distribution of an ESC, however, is not likely assuming that convergent evolution of exactly the same ESC is improbable and given Postulate #1 stipulating mutational connectivity among ESC character states. Therefore, distribution of ESCs among species should contain a strong phylogenetic signal.

The phenotypic stability promoted by an ESC suggests that, at the appropriate level of description, ESC characters will exhibit stasis (the failure to evolve under a null model of evolutionary change by whatever means—mutation, selection, drift—given a long enough period of time; Schwenk, 1995b). Stasis at the character level implies retention of the ancestral state, or plesiomorphy (Fig. 4). Stasis in a functional system will therefore be represented in a comparative analysis as a pattern of symplesiomorphy in an array of characters drawn from that system, in contrast to non-ESC characters, which would be variously apomorphic and plesiomorphic (Fig. 5). Homologous (ESC) characters in related, non-ESC taxa are expected to manifest variable, derived states owing to the fact that the characters in these taxa are "free" to evolve, although there is no reason why some of them might not be plesiomorphic, either through reversal, or because they are otherwise stabilized (e.g., constrained or maintained by external selection), therefore retaining the ancestral condition.

By "appropriate level of description" we mean representation of a character by its fundamental attributes, that is, those qualities of the character that reflect a particular design principle or its role in maintaining the proper function. ESC characters described in this way represent the "diagnostic set" referred to previously, which is a character description of the ESC core. However, subsidiary states of core characters (i.e., variants that are subsumed within the more general core description) might vary among species exhibiting the ESC, but only within phenotypic parameters set by the proper function (Table IV, Figs. 3, 6). For example, a diagnostic character of the Case Study ESC is possession of long, deep ("high-profile") foretongue papillae. Among ESC species, however, high-profile papillae can be either "filamentous" or "reticular" in form (Schwenk, 1988). Both papillary types perform putatively identical roles in lingual prey capture, the proper function of the ESC. They maintain functional equivalency owing to their high-profile form, which is a critical part of the lingual adhesive mechanism (Schwenk, unpublished data). Therefore, despite significant morphological differences, both papillary types embody the same fundamental design principle. In contrast, non-ESC species exhibit variable, "low profile" foretongue papillae (Schwenk, 1988) that violate the central design principle of the ESC by failing to maintain functionality of the lingual adhesive mechanism (Tables I, IV; Figs. 3, 6). Thus, it is ESC characters defined at their

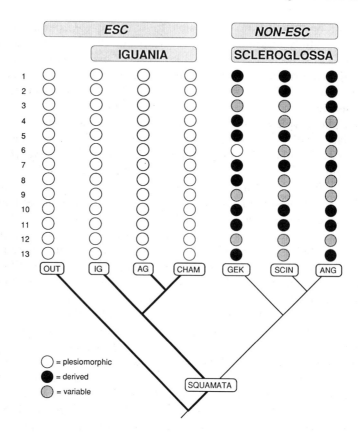

FIG. 5. Phylogenetic signature of an ESC. At the appropriate level of description, the ESC characters remain invariant among all species that maintain the ESC, in this case *Sphenodon* as outgroup and all species of the Iguania. The characters are numbered according to Table I. Note that the characters are variable among the Scleroglossa, which has evolved a new form of prey prehension using its jaws. AG = Agamidae; ANG = Anguimorpha; CHAM = Chamaeleonidae; GEK = Gekkota; IG = Iguanidae; OUT = outgroup; SCIN = Scincomorpha.

diagnostic level that exhibit a pattern of symplesiomorphy across a phylogeny. Character states nested within the diagnostic set might be variously derived (e.g., within high profile there are filamentous and reticular papillae, as noted previously). Variability in character states at this lower level of description allows for within-ESC phylogenetic patterns in character evolution, discussed in Appendix 2.

The presence of an ESC might generate a consistent pattern of symplesiomorphy in at least two different sets of characters: the diagnostic set,

TABLE IV.   Within-ESC Character Variation in Iguania

Character	Character States
ESC Characters	
1. foretongue papillae (Table I, char. 1)	a. filamentous b. reticular
2. foretongue glands (Table I, char. 4)	a. papillary surfaces only b. mucous crypts penetrate into tongue musculature
3. foretongue surface epithelium (Table I, char. 5)	a. stratified squamous b. arborate c. both present
4. lingual process form (Table I, char. 6)	a. tapered b. parallel-sided
Autapomorphies of Chamaeleonidae	
5. ballistic phase of SO II	a. absent b. present
6. posterior lingual limbs	a. well-developed b. reduced/absent
7. genioglossus insertion	a. side and ventral surfaces of tongue b. lingual sheath
8. hyoglossus form	a. longitudinal bundle b. elongated, folded
9. hyoglossus sarcomeres	a. typical b. supercontracting
10. verticalis muscle	a. typical b. modified into hydrostatic accelerator muscle
11. laryngohyoid ligament	a. present b. absent
Related, Non-ESC Characters	
12. verticalis fiber orientation in hindtongue	a. vertical b. circular c. radial
13. genioglossus internus	a. absent b. present
14. dentition	a. pleurodont b. acrodont

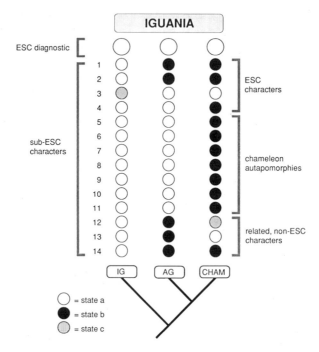

FIG. 6. Within-ESC character variation. These characters (see Table IV) are predicated on the ESC but do not belong to its functional core. Note that characters vary and show evolutionary trends, such as the acquisition of chameleon autapomorphies, but they do not interfere with the integration of the ESC core characters. States a, b, and c correspond to those in Table IV. AG = Agamidae; IG = Iguanidae; CHAM = Chamaeleonidae.

as described previously, and sets of characters drawn from functional systems not part of the ESC, but functionally related to it. In the Case Study, for example, some characters of the vomeronasal chemosensory system are shared by the ESC system owing to common use of the tongue. As argued previously, internal selection within the ESC limits evolution in tongue phenotype to character states suboptimal for chemosensory function (a functional trade-off). In addition, functional dependence of other (non-ESC) vomeronasal system characters on tongue phenotype results in an evolutionary cascade such that phenotypic evolution of the iguanian vomeronasal system, as a whole, is also limited. Comparative analysis of chemosensory characters reveals a parallel pattern of symplesiomorphy. In fact, it was this phylogenetic pattern that first suggested the action of some form of putative constraint limiting adaptation of the vomeronasal system within Iguania (Schwenk, 1993, 1995a). With disruption of the ESC in the ancestors of Scleroglossa, the functional coupling between the feeding (ESC) and vomeronasal systems (stemming from functional trade-offs of a shared subset of characters) was broken. Release of the tongue from mechanical demands of prey prehension permitted it to respond to preexisting selection for chemosensory function, initiating adaptive optimization of many other (nonlingual) vomeronasal characters in some scleroglossan taxa. In general, morphologists have observed that "couplings" (both anatomical and functional) tend to limit phenotypic evolution and that the "decoupling" of previously coupled parts/systems can trigger subsequent morphological diversification within a lineage (e.g., Vermeij, 1974; Lauder and Liem, 1989; Roth and Wake, 1989; Galis, 1996; Galis and Drucker, 1996; Schaefer and Lauder, 1996).

Finally, we note that a phylogenetic pattern of symplesiomorphy in a set of characters drawn from a given functional system is not sufficient grounds for diagnosing an ESC. As with constraint, a phylogenetic pattern of stasis is a good starting point for beginning investigation, but it fails to exclude other conditions that might lead to a similar pattern (Schwenk, 1995a). A similar phylogenetic pattern in character evolution could be caused by stabilizing external selection (in the traditional sense of constancy in external environment); functional coupling of a system to an ESC. or otherwise stabilized system (as noted previously); or genetic and/or developmental restrictions on character variation, that is, developmental constraint in the strict sense. At the least, a convincing ESC diagnosis will pair comparative character analysis with a functional analysis detailing the web of integration among the characters, thus plausibly linking pattern and causality.

Patterns of Character Evolution within an ESC

As long as proper function is maintained, ESC characters are free to vary within certain design parameters, thus permitting adaptation within the ESC and the attainment of locally optimal phenotypes (Table IV; Figs. 3, 6). Adaptively neutral character evolution is also possible.

Given within-ESC character evolution and therefore the possibility of distinct, sub-ESC phenotypes, why is the ESC concept not applied to these lower hierarchical levels? First, the ESC is defined in terms of a given proper function and thus, the concept minimally applies to the largest set of (mutationally connected) phenotypes maintaining the proper function. For example, although there are distinct lingual-feeding phenotypes within the Iguania (Case Study), the ESC applies to the entire clade because all maintain the proper function of lingual prey capture; alternative phenotypes among ESC-containing taxa are seen to be "variations on a theme" (they maintain a fundamental design principle). Second, sub-ESC phenotypes are varyingly stable and possibly transitory. To the extent that they represent local, adaptive optima, they respond to external selection and do not exhibit the stability inherent to the ESC core characteristics. Once again, the ESC concept is dependent on recognition of the hierarchical level at which self-stabilization occurs and it is at this "diagnostic," or core level that long-term evolutionary stability is expected.

There are two possible patterns of within-ESC character state evolution: either ESC character states are distributed among ESC-containing taxa such that they contain a phylogenetic signal (i.e., are correlated with evolutionary history), or their pattern of distribution is phylogenetically uninformative. For each of these modal patterns there are several potential explanations. Although analysis of such patterns is relevant to understanding the limits to character evolution within ESCs, it is peripheral to our arguments here, hence we relegate this discussion to Appendix 2.

Phylogenetic Conclusions

Overall, comparative methods offer only limited help in identification and diagnosis of an ESC owing to the range of potential variation in within-ESC character evolution. The strongest phylogenetic pattern predicted is in the representation of an ESC across a phylogeny and the expectation of monophyly or paraphyly of ESC-containing taxa. At the character level, a pattern of symplesiomorphy in ESC character states is expected in diagnostic characters and therefore cladistic analysis is an important component of ESC diagnosis (Figs. 5, 7). Comparative analyses that reveal patterns of evolutionary stasis in functionally related characters should lead to questions about processes underlying the pattern; presence of an ESC is only one plausible explanation for such a pattern. Other possible explanations are noted previously.

A prediction emerging from our analysis is that diagnostic character phenotypes should be buffered from environmental change and therefore should be largely independent of ecological factors. A useful approach to testing this prediction is the application of Felsenstein's (1985) method of independent contrasts to the diagnostic character set (there are currently a variety of techniques and software packages that adopt this methodology; e.g., Harvey and Pagel, 1991; W. P. Maddison and Maddison, 1992; Garland, 1992; Garland et al., 1992; Martins, 1996a, 1996b; Martins and Hansen, 1997). Such methods assess patterns of correlated character state change within and among clades. If one includes both phenotypic and environmental char-

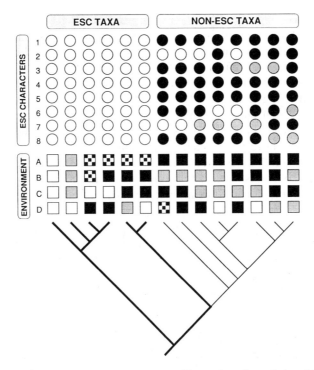

FIG. 7. A hypothetical comparative analysis illustrating the relationship between ESC character variation and the environment. 1–8 represent ESC core characters and A–D represent four environmental parameters (e.g., available prey types, predators, amount of cover, annual rainfall, etc.). Different fillings and patterns indicate different character (or environmental) states. Note that ESC characters remain uniformly stabilized despite extensive variation in aspects of the environment that are expected to exert selection pressures on these characters. In taxa that have escaped from the ESC phenotype, character evolution is variable. Furthermore, in non-ESC taxa, variation in some characters (1, 3, and 7) is precisely correlated with certain features of the environment (A, B, and C, respectively), suggesting the likelihood of adaptive matching between these phenotypic traits and the changing environment mediated by external selection. Phylogenetic concordance between loss of the ESC and a change of state in environmental parameter A suggests the possibility that a shift in this aspect of the environment might have played a causal role in the disruption of the ESC.

acters in the analysis, then one can reveal the extent of correlation between habitat shifts (changes in habitat character state) and phenotypic evolution (changes in phenotypic character state). Presence of an ESC should result in little or no correlation between (diagnostic) character phenotype and environmental change. Implicit is a null hypothesis of responsiveness of phenotypic traits to external selection pressures, that is, a lack of internal buffering. If environmental parameters are invariant for all ESC-containing taxa, then the ESC hypothesis cannot be tested. In other words, phenotypic stability promoted by intrinsic factors, such as functional integration, cannot be distinguished from maintenance of the phenotype through a stable external selection regime (i.e., "phylogenetic niche conservatism"; Harvey and Pagel, 1991). However, phylogenetic (or fossil) evidence of very long-term stability of the putative ESC phenotype increases the likelihood that environmental change has transpired historically, even if current environments are uniform, suggesting a role for intrinsic, organismal attributes in resisting phenotypic change.

Diagnosis and description of an ESC through independent contrasts methods, of course, is subject to limitations of the methods generally; for example, the need for a well-supported cladogram inclusive of all relevant taxa, in most cases the necessity of known or estimated branch lengths, restriction to either discrete or continuous characters, and the need to assume a particular null model of evolution (e.g., Brownian motion) that might be unrealistic for the particular characters in question (Harvey and Pagel, 1991; Díaz-Uriarte and Garland, 1996; Hansen and Martins, 1996).

In concluding this section we emphasize that cladistic analyses revealing symplesiomorphy and stasis and independent contrasts analyses suggesting lack of correlation between phenotypic and environmental parameters can only provide suggestive patterns. Although the patterns can help to eliminate some alternate models of phenotypic evolution, they must ultimately be linked to an hypothesis of causality (of which presence of an ESC is but one) if we are to address the fundamental questions of phenotypic evolution. In linking pattern to process, we must rely on those emergent properties of living systems only evident in whole organisms (epigenetics, development, function, behavior, phenotypic plasticity) and assess the extent to which these processes could generate the patterns observed. In other words, emergent organismal attributes (such as functional integration) can influence the direction and nature of unit character evolution, resulting in the generation of particular patterns. It is in the "reciprocal illumination" (Hennig, 1966) of character analysis and organismal biology that elucidation of ESCs specifically and phenotypic evolution generally can be sought.

A BRIEF REVIEW OF RELEVANT CONCEPTS: FUNCTIONAL INTEGRATION AND INTERNAL SELECTION

Many of the ideas expressed in this chapter are very old and have appeared in the literature in various forms. These ideas reflect the growing sense that neo-Darwinian theory is not complete because it fails to embrace emergent properties of whole organisms, including temporal and functional interactions among component parts, in its approach to phenotypic evolution. The notions of internal selection and functional integration are a critical part of this *Zeitgeist* and the ESC model emerges as a formalization of these oncepts. An exhaustive historical review is beyond the scope of this chapter, but we highlight here a few important examples.

Evolutionists have noted the existence of persistent morphological forms at least since Darwin (1859), who recognized that some lineages evolved morphologically at a very slow rate (he cited the horseshoe crab, *Lingula*, as an example). Unsurprisingly, Darwin attributed evolutionary rate differences to differences in the external environment (physical factors, competition, marine vs. terrestrial, etc.). However, he also noted as part of his explanation that lineages could vary in the degree to which they accumulate phenotypic variation upon which selection could act (i.e., lineages differ in phenotypic "variability," *sensu* Wagner and Altenberg, 1996), thus opening the door to (but never entering) the possibility of intrinsic, organismal determinants of evolvability.

Several late 19th and early 20th century biologists explored the notion of intrinsic components of selection, notably Roux (1881) and Weismann (1896, in Whyte, 1965). These authors assigned Darwinian selection to cells and to organs, seeing in their growth and multiplication an analogy for the competition among individuals so central to Darwin's theory of natural selection. Weismann called such intraorganismal competition "intraselection." However, as Whyte (1965) pointed out (see the following paragraph), intraselection is distinct from internal selection because it arises from competition among parts, rather than the coordination of function among parts. Intraselection was discredited by later workers who showed that tissues and organs develop autonomously to programmed limits without competition for internal resources, but recent empirical work with insects has resurrected the idea, showing that it may be a valid model in some cases (Klingenberg & Nijhout, 1998; Nijhout & Emlen, 1998).

Following these early references to intrinsic forms of selection there was increasing consideration of the possibility that organismal attributes, and not just the external environment, contribute to directions and patterns of phenotypic change. These are reviewed by Whyte (1965). We note two important cases here, however. A central element of Schmalhausen's (1949) theory of stabilizing selection was that as lineages evolve their developmental systems become increasingly "autonomous," by implication buffered from external influences. Schmalhausen viewed developmental systems as relatively plastic in more "primitive" groups and more internally stabilized (we might say canalized) in more derived groups. Schmalhausen's writing is somewhat obscure, so we would not want to attribute more to this than is warranted, but his ideas do suggest the notion that developmental characters are initially sensitive to external selection, but that subsequent integration renders them more internally selected. Schmalhausen was not, however, concerned with functional integration, per se. Several years later, von Bertalanffy (1952) returned to the earlier idea of intraselection but extended the notion to multiple hierarchical levels and to the organizational properties of organisms, that is, functional integration or internal selection (see quote in Whyte, 1965, p. 82).

Although we have used a dialectical approach here, tending to contrast neo-Darwinian thinking with internalist (or structuralist) approaches to evolution, the "synthesis" was not without structuralist elements. A notable thinker along these lines was G. G. Simpson. Simpson (1953), like Darwin, saw persistent forms as a manifestation of slow evolutionary rates, which he called "bradytely." Also like Darwin, he was concerned with whole-organism phenotype and not parts of organisms, or systems, as we are here, but his comments are nonetheless applicable. Simpson's view of phenotypic stability was somewhat more complex than Darwin's, however, and bears scrutiny. Because some bradytelic lineages include phenotypes that have persisted over the long term through multiple environmental transitions, Simpson recognized that bradytely failed to conform to the implicit neo-Darwinian expectation of adaptation through form–environment matching. He therefore sought an explanation for this phenomenon, but concluded that it was fully compatible with his stolidly neo-Darwinian view of adaptive evolution:

> Evolutionary change is so nearly the universal rule that a state of motion is, figuratively, normal in evolving populations. The state of rest, as in bradytely, is the exception and it seems that some restraint or force must be required to maintain it. This force undoubtedly exists and is the same as the force that usually orients evolutionary change: selection. (p. 331)

However, Simpson considered selection to occur at four environmental levels, the last of which he called the "individual (internal) environment," noting that the "distinction between inside and outside is by no means as clear as at first appears" (Simpson, 1953, pp. 162–163). He further expressed tentatively the notion of functional integration and internal selection, as we conceive it here, in stating, "Each individual has its own internal environment to which it is also adapted and *within which its various parts and functions are correlatively adapted*" (Simpson, 1953, p. 162; italics added). In formulating these opinions, Simpson credited the works of Sinnott (1946) and especially Emerson (1949, in Allee *et al.*, 1949), who made a similar distinction between "exoadaptation" and "endoadaptation" (Allee *et al.*, 1949, p. 631). Emerson (1949, in Allee *et al.*, 1949) was particularly clear on this point: "The existence of complex internal adaptation between parts of an organism or population, with division of labor and integration within the whole system, is explicable only through the action of selection upon whole units from the lowest to the highest" (p. 684). Thus, Simpson's notion of natural selection included within it the kernel of internal selection, so that when he stated that selection actively maintains stable phenotypes, he potentially allowed for the action of internal selection. However, in emphasizing that there is no crisp distinction between internal and external environments (Sinnott's [1946] view), Simpson emphasized external, environmental factors in adaptive evolution, never fully considering the possibility that internal selection and external selection are qualitatively different phenomena that could act *in opposition* to one another, possibilities we advocate here (also Whyte, 1965; Arthur, 1997; see the text that follows).

Although he did not explicitly consider the role of internal selection in promoting phenotypic stability, Simpson (1953) did imply that bradytelic lineages possessed intrinsic, organismic properties that limited the rate at which they evolved. He specifically addressed the possibility that "certain genetic and other factors *in organisms*" (p. 327, italics added), including physiological and morphological traits, might limit the rate of phenotypic evolution, but he dismissed these for lack of supporting data.

Simpson's (1953) conclusions are ultimately unsatisfying because despite his general acceptance of a role for the internal environment in natural selection and his insistence that organisms bring with them to the evolutionary arena their own, historically determined (intrinsic) attributes, Simpson finally stressed the role of stable, external environments and the action of stabilizing ("centripetal") selection in the maintenance of phenotypes: "As to the environment, the principal requirement (for arrested evolution) is that it persist and do so with less fluctuation than the maximum that can be met by existing and expressed, not potential, variation in the populations concerned" (p. 333). Slowly evolving, or static groups, were believed to occupy a "stable and persistent zone" characterized by "a long, eventless course of relatively unchanging conditions" (p. 331).

Dullemeijer (e.g., 1956, 1959, 1974, 1980, 1989) has been a forceful advocate of the view that form can only be understood in the context of morphological and functional integration among component parts, a view heavily influenced by his predecessor at Leiden, van der Klaauw (e.g., 1945). He viewed each individual morphological element as the center of a "field of decreasing . . . constructional consequences" (Dullemeijer, 1980, p. 226), which he called a "functional component." Organisms, therefore, comprise numerous, overlapping functional components that exert multiple, intrinsic demands on the form of their central elements. Dullemeijer thus encapsulated the notion of internal, functional (and anatomical) integration, which he acknowledged must influence patterns of phenotypic evolution. However, functional components center around individualized phenotypic elements (what we might call characters) and do not necessarily correspond to functional systems, though Dullemeijer stressed that systems are themselves composed of functional components.

Many others have suggested, in one form or another, a role for functional integration in phenotypic evolution. The notion of "functional units" or "complexes" has a long history in vertebrate morphology starting with van der Klaauw (1945). Bock and von Wahlert (1965) and Bock (1964), for example, considered functionally related character sets to represent "functional units" that might behave evolutionarily as single characters. Bock and von Wahlert (1965) noted that "The adaptiveness of a character complex is usually dependent upon a close correspondence in the form and function of its individual features" (p. 272), seeming to imply the action of internal selection in shaping the form of individual characters. Bock (1963) further suggested that morphological evolution might be limited in some groups "because the features of the adult as well as in the developing embryo must be in structural and functional harmony at all times" (p. 280). However, Bock & von Wahlert (1965) ultimately viewed characters within functional units as readily influenced by environmental selection and the units themselves as potentially labile. Gans (1969b) pointed out that what are usually described as "functional units" or "components" are, in fact, topographic or anatomical regions that he preferred to call "mechanical units" in the absence of functional data relating how component parts are integrated. He further noted that a true functional unit would include not only its musculoskeletal components, as usually described, but other functionally related elements of the system. Riedl's (1978) concept of burden (discussed previously in "Origin of an ESC") has functional integration implicit within it, although burden especially implies temporal integration through ontogeny. The Leiden school of "transformation morphology," heavily influenced by Dullemeijer, has continued to emphasize the importance of anatomical and functional relationships among parts in influencing pathways of evolutionary change (e.g., Barel, 1993; Galis, 1996). Rollo (1995) considered integration at all hierarchical levels to be of overriding importance in phenotypic evolution and regarded integration and coadaptation among parts and hierarchical levels to be a central issue in evolutionary biology that had only rarely been specifically addressed. Rollo (1995) called functionally integrated systems "adaptive suites" in which "each feature of the phenotype may be adaptively honed by evolution for maximal competence, but the integration of various aspects may be of even greater importance" (p. 256).

He proposed that higher order functional integration imposes constraints on genetic structure such that coadapted genomes result, further stabilizing the phenotype. Similarly, Hall (1996) recognized that functional integration could limit phenotypic evolution, a condition he called "functional constraint" (see the following section).

Of greatest relevance to the present work is the important, but almost totally neglected, book of Whyte (1965), an explication of several earlier papers. Whyte formally coined the term "internal selection" (although he attributes its earlier use, in a narrow sense, to C. Stern) by which he meant "the restriction of the directions of evolutionary change by internal organizational factors" (pp. 57–58). Whyte advocated a structuralist approach emphasizing not the atomized parts constituting an organism but the ordered relationships among the parts evident at every hierarchical level, which he referred to as the "coordinative conditions (or C. C.)." (We would describe such conditions as "emergent properties" of organisms.) Coordinative conditions arise from the spatio–temporal coordination of the parts within an organism expressing "the general conditions of biological coordination; the rules of geometrical and kinematic ordering which must be satisfied (to within a threshold) by the internal parts and processes of any viable cellular organism" (p. 7). Most important, Whyte showed that internal selection is an organismal attribute clearly distinguishable from selection imposed by external, environmental conditions; for example:

> It is good style today to equate the internal and external environments; therefore there can be only one universal type of natural selection. Thus contemporary habit tends to inhibit the recognition of a distinction which is indispensable to clear thinking. For the internal and external environments are not alike; life is not a vague property diffused throughout the universe; the nature of life lies partly in the fact that it is concentrated in units whose interior is subject to an ordering principle which does not apply to the inanimate spaces between organisms. These units of organization do not lose their distinguishing form of order because they are in perpetual give and take with the environment; the C.C. do not operate through open spaces, they are only effective inside boundaries. Thus the two kinds of selection can and should be distinguished. (pp. 58–59)

Whyte (1965) encapsulated the essence of phenotypic stability arising from internal selection by pointing out that the organizational properties of living systems limit the potential range of evolutionary pathways from a given starting point. He acknowledged that specific adaptations arise by Darwinian, external selection, but only within the "broad avenues determined by [an organism's] own structural nature" (p. 53), an idea we formalize here in our exploration of sub-ESC evolution (Appendix 2). Finally, it is important to note that Whyte emphasized in his conception of internal selection lower hierarchical levels, particularly the role of molecular, chromosomal, and cellular levels in "screening" mutations before they are subject to external selection. Functional integration at all levels, however, is implicit in his discussion.

Recently Arthur (1997) has resurrected and incisively amplified Whyte's (1965) notion of internal selection. What's important, Arthur has more explicitly included functional integration within its fold and has framed internal selection more explicitly in the context of neo-Darwinian theory, as we attempt here. For example, Arthur redefined internal selection in terms of fitness: "Internal selection occurs when individuals with different genotypes at a given locus differ in fitness because they differ in their degree of internal coordination" (p. 221). Arthur further developed the contrast between "positive" and "negative" internal selection. The former is more controversial in our view; it implies an important role for internal selection (in combination with external selection) in driving phenotypic reorganizations at the inception of new forms. Arthur suggested that the stabilizing role of negative internal selection followed on the creation of a new phenotype, acting to maintain its functionality by limiting variation. Our emphasis here has been on Arthur's (and Whyte's) sense of negative internal selection, that is, a form of stabilizing selection that promotes long-term phenotypic stability.

Arthur (1997) proffered a series of statements about internal selection that parallel our postulates in purpose and in some cases, meaning. For example, he pointed out the environmental independence of internal selection and used it to predict its role in phenotypic stability. He predicted, as do we, epistatic effects on fitness. However, like Whyte (1965), Arthur's emphasis was primarily the temporal integration of development and its implications for phenotypic evolution. We note that owing to the long gestation of this chapter we arrived at our conclusions, postulates, and definition independent of Arthur. We take such confluence as evidence that Whyte's ideas are timely in the current *Zeitgeist*, premature though they were in his own time.

Others have independently elaborated similar notions of internal selection. Dullemeijer, (1980), for example, arrived at "internal selection" based largely on Waddington's (1957) theory of canalizing selection, apparently unaware of Whyte's (1965) earlier use of the term: "... for each element in an organism the others form its environment. Thus if such an interaction selects individuals and contributes to genetic change, this selection is natural but internal" (Dullemeijer, 1980, p. 196). Dullemeijer (1980) concluded that internal selection should be distinguished from natural selection because it lacks an environmental factor. He also recognized that integrated systems are to some extent buffered from the effects of environmental selection and that they contribute to phenotypic stability.

Wake *et al.* (1983) and Roth and Wake (1985) proposed a holistic model of organisms as "autopoietic" systems, that is, they are self-maintaining and internally stabilized. Their model was applied to whole organisms rather than to functional systems within an organism. Wake *et al.* (1983) suggested that within organisms "a network of interaction exists which constitutes a circularly closed system" (p. 218). Their organismal-level view is similar to Dullemeijer's (1974, 1980) "functional component," but differs importantly in its emphasis on functional interactions and explicit relationship to evolutionary change:

> The range of ontogenetic and phylogenetic change of one element is, therefore, determined by the structural and functional properties of all other elements. Each ontogenetic or phylogenetic change of the system must remain within the functional limits of the process of circular production and maintenance of the elements, or the system itself will decompose (Wake *et al.*, 1983, p. 218).

Wake *et al.* (1983) further noted that environmental (external) selection acts on the system, as a whole, and that external influences on individual elements are limited by the effects of potential changes on system function, thereby implying the buffering capacity of functional integration on external selection. Clearly evident in this view is the dichotomy of internal versus external selection and the stabilizing effect of "internal dynamics," including developmental constraint and internal selection, on phenotypic evolution.

DEVELOPMENTAL AND FUNCTIONAL CONSTRAINT

Phenotypic stability is often attributed to constraint, that is, the notion that some factor is limiting the ability of a phenotype to evolve (see overviews by Alberch, 1982; Hall, 1996; Maynard-Smith *et al.*, 1985; McKitrick, 1993; Schwenk, 1995a; Schlichting & Pigliucci, 1998). Indeed, a pattern of evolutionary stasis alone is often taken as *prima facie* evidence of constraint (Schwenk, 1995a). However, in many cases stasis can be equally, or more plausibly, attributed to stabilizing selection (e.g., Hall, 1996; Schwenk, 1995a); as such, phenotypes are *maintained* in their plesiomorphic states by the action of natural selection, as opposed to being *unable to respond to selection* owing to some intrinsic limitation on production of phenotypic variation (such as lack of genetic variation or a highly canalized developmental system). Some workers have actually equated stabilizing selection with constraint ("selective constraints") or have failed to recognize that an identified constraint was, in fact, a form of selection (Schwenk, 1995a).

The conflation of constraint and selection has been a major factor in the dilution of the constraint concept (Antonovics and Tienderen, 1991; Schwenk, 1995a; Pigliucci *et al.*, 1996). However, to a large extent this dilution has occurred because of the general failure to distinguish between external and internal selection. The issue of constraint and its relation to natural selection is sufficiently complex that it requires a separate treatment beyond the scope of this paper (Schwenk and Wagner, in preparation). However, we note that two broad categories of constraint can be recognized: developmental constraint and functional constraint. Functional constraint is particularly relevant to the ESC concept because it nearly always implies limitation in character evolution owing to the functional integration of the character with others (Hall, 1996). As such, functional constraint implies the action of internal selection. The ESC concept could, therefore, be cast as a model of functional constraint. Functional trade-offs constitute another mode of internal selection leading to phenotypic stability or functional constraint.

Developmental constraint, on the other hand, has a far more complex relationship to natural selection that requires additional context (Schwenk and Wagner, in preparation). In general, developmental constraints imply restrictions on the availability of phenotypic variants on which current selection can act, thereby limiting the action of selection. One type of generally accepted developmental constraint is limitation in the generative systems that build phenotypic characters (e.g., Alberch, 1982; Maynard-Smith *et al.*, 1985; Wagner, 1994). Developmental constraints, in this sense, are relevant to the ESC concept indirectly. As noted in the first postulate (as discussed previously), an ESC is conceptualized as a cluster of phenotypic states that are mutationally connected; that is, one can reach each member of the ESC from another member by mutation without having to go through states that do not belong to the ESC. In addition, it is assumed that it is not possible to jump directly from one ESC to another by a single mutation, as from rodent to ungulate type of mastication. This postulate then makes assumptions about which phenotypic transitions are possible by a single mutation and which ones are not; that is, it implicitly makes assumptions about the existence of developmental constraints. However, the role that developmental constraints play in the ESC concept is different from the way the concept is used in other contexts. For the ESC concept it is only necessary to assume that one cannot realize all phenotypic transformations by a single mutation. If this is the case it implies something like a "mutational topology" on the set of possible phenotypic states, that is, the notion of neighborhood and distance in the space of possible phenotypes (Jones, 1995).

PROBLEMS

Given the complexity of the situation that the ESC concept is expected to capture, the list of open problems has to be longer than this already long chapter. However, we believe that the principal open problem is one of methodology: What are the criteria to assess whether a conservative phenotypic system is an ESC in the sense of this chapter? Do ESCs in fact exist?

This problem has been to some extent discussed in the section about the expected phylogenetic signature of an ESC. However, it was also noted that the comparative method alone can only be a heuristic, which may point to some candidate ESCs, because the expected phylogenetic pattern *per se* is not different from that expected under some other causal scenarios. Hence, attempts to test whether a certain taxon possesses an ESC have to be two-pronged. On the one hand, one has to determine whether alternate explanations, like developmental constraint and external stabilizing selection, can be rejected. On the other hand, one has to test each of the five postulates that were given to characterize the ESC concept. Most of them are statements about the functional relationships among the parts of the ESC and thus have to be tested by functional analysis. However, the functional analysis has to be done in a phylogenetic framework where the question is whether the functional integration and uniqueness applies to all members of a mono- or a paraphyletic taxon.

Perhaps the most critical test of an ESC hypothesis is possible when the taxon that exhibits the putative ESC is paraphyletic and the derived clade does not have the ESC, as reviewed in the Case Study. The predictions then are that these escapes are rare (a phylogenetic hypothesis) and that the transition from the putative ESC to the alternate state has not been caused by direct selection for the proper function of the ESC but by one of the scenarios elaborated in the section titled "Escape from an ESC." Whether it is possible to determine this, of course, depends on the amount of information available about the particular taxon. Diverse and well-studied animal groups, such as the lungless salamanders (Wake and Larson, 1987; Roth and Wake, 1985; Wake, 1991), provide the opportunity to test such historical scenarios.

SOME GENERAL IMPLICATIONS OF THE ESC CONCEPT

Evolutionarily conservative patterns of the phenotype occur on various levels of biological organization, from deeply entrenched properties like the genetic code to whole-body plans. How does the ESC concept fit into this context? We believe that there are roughly three levels of organization where stable biological entities occur: characters, character complexes (ESCs), and body plans (*Baupläne*). The emergent stability evident at each level is presumed to reflect different causal mechanisms. As such, each entity defines a different category of natural units, each requiring its own empirical and conceptual approach not reducible to (even if informed by) the study of any other-level units.

Morphological characters have traditionally been recognized as homologs, that is, individualized parts of the body with conserved phenotypic features (Wagner, 1989; Hall, 1992). Homologs are considered to be stable in spite of changes in function (e.g., tetrapod limbs used for walking, for swimming, or for flying). Therefore it has been hypothesized that the conservative nature of characters is due to developmental rather than functional constraints (Wagner, 1986; see Schwenk and Wagner, in preparation, for a discussion of constraints).

ESCs, as conceptualized here, are composed of characters and are stable because of the functional relationships among them. Their stability is thus predicated on the maintenance of a proper function as opposed to stability *in spite of* functional change, as is the case for characters. ESCs thus imply constraints on evolutionary change that go beyond those implied by developmental constraints.

At the highest level of the organismal hierarchy are *Baupläne*, the theoretical status of which is currently under debate (Hall, 1992, 1996; Arthur, 1997; Gerhart and Kirschner, 1997). The apparent stability of *Baupläne* seems to be caused by both developmental and functional factors, because they determine the general developmental context in which characters develop (e.g., whether the animal is diplo- or triploplastic) and the functional matrix in which the characters are integrated. We therefore resist the temptation to call a *Bauplan* the most inclusive ESC (but see Wake *et al.*, 1983).

From this outline it should be clear that we consider the study of functionally integrated character complexes (ESCs) a specifically circumscribed approach to organismal biology aiming at those patterns that are explicable by functional integration. We do *not* suppose that the ESC concept in itself can give rise to a general theory of biological organization, but we do believe that the study of functionally integrated character complexes is a paradigm for the necessity and heuristic power of phenotype-level approaches in biology.

ACKNOWLEDGMENTS

We thank the members of YUCEE (Yale-UConn Evolution Ensemble) for discussion, and Frietson Galis, Kent Holsinger, Ken Kardong, Junhyong Kim, Mark McPeek, Dan McShea, Sean Rice, Carl Schlichting, Marvalee Wake, and David Wake for critically reading the manuscript. Frietson Galis shared her extensive knowledge of fish functional morphology and unpublished research results. The financial support of the National Science Foundation (BIR-9400642 to GPW, IBN-9601173 to KS), the Yale Institute for Biospheric Studies, and the University of Connecticut Research Foundation, is gratefully acknowledged. This is contribution number 52 of the Center for Computational Ecology at Yale.

APPENDIX 1: MILLIKAN'S (1984) "PROPER FUNCTION"

The concept of biological function is highly complicated and multifaceted. An excellent introduction to the conceptual problems associated with the notion of "function" is provided by Chapter 1 of Millikan's (1984) monograph. The solution proposed by Millikan is her concept of proper function. The concept of proper function is complementary to the current usage of the term "adaptation" as a character that was shaped in order to perform a certain function that contributes to fitness (Rose and Lauder, 1996). In short, a proper function is what an adaptive character is adapted to perform. In other words, the proper function is the reason for the existence of the character. Of course the theory of proper functions is much more subtle and general than this short characterization. The reason is that Millikan developed this concept as a device to explain the structure and function of language by considering language as just another instance of a historically derived characteristic of living organisms.

There has been legitimate criticism of the usefulness of the proper function concept (e.g., see Amundson and Lauder, 1994), particularly if proper function is considered the only defensible concept. We agree with these authors that the proper function concept is inadequate in those disciplines of biology that focus primarily on the functioning of organisms (e.g., functional morphology or physiology) rather than studying the evolutionary origin of adaptations. A "causal role" definition of function is better suited to these disciplines (Amundson and Lauder, 1994). In the present context, however, we are interested in the phylogenetic modification of functional parts of organisms rather than their physical actions. We therefore maintain that proper function is the appropriate concept to explicate the ESC concept.

An example of a conceptual problem solved by the notion of proper function is the question of whether a character itself actually needs to perform the function in order to "have" this function. For instance, only a tiny fraction of all sperm cells actually fertilize an egg cell. Does this mean that only those that actually penetrate the egg membrane have this function, but not the others? This is not the usual meaning of the word "function." Actual participation in a process is not necessary for a structure to have the function. Hence proximate involvement is not necessary; function refers to a process of historical continuity mediated through a mechanism of inheritance.

The only obvious hole in the definition of proper function by Millikan (1984) is the question of what she means by "*m* performs a function *F*." A possible solution to this problem is to say that a character performs a function if it mediates a causal relationship between two events. For example, hearts mediate a causal relationship between biological energy (i.e., ATP or glucose) and the flow of blood. There is no inherently necessary, physical relationship between ATP and blood flow. For the energy stored in ATP to be transformed into the kinetic energy of a fluid, a mediating device is necessary that first turns ATP into mechanical action, and second, turns mechanical action into volume changes and a directed blood flow. To locally provide this causal relationship is what hearts usually do.

We think that the notion that *performance means mediating causality*, where pure physics does not provide any, can be generalized to all instances where the concepts of performance and function are appropriate. Language mediates a fairly regular relationship between speakers' intentions and reactions of the listeners/readers. Language creates locally causal relationships within a given communication system (community of language users).

APPENDIX 2: CHARACTER EVOLUTION WITHIN AN ESC

No Phylogenetic Signal

1. *Lack of variable states*. It is theoretically possible that within-ESC characters do not vary. As such, diagnostic character set descriptions are adequate to encompass total character variability. In other words, all ESC characters exhibit a single state in all ESC taxa and the ESC is represented by a single phenotype rather than a set of phenotypes. Such a pattern would suggest a very recent origin of the ESC and/or that internal selection on character phenotype is so severe that variants are not tolerated. A third, unlikely possibility is that ESC characters are subject to generative (genetic/developmental) constraints that prevent the origin of novel states (see the following paragraph). In any case, such a pattern, if it occurred at all, would be most likely to occur in an ESC with very few characters.

2. *Isotropy of ESC character space*. In the purest conception of an ESC, mutational connectivity of variable character states implies isotropy in state transitions; that is, any state of a variable character can be transformed into any other state without directional bias or fitness consequence. An implication of this condition for comparative analyses is that homoplasy is expected to be commonplace in ESC characters if the lineage is sufficiently old. As such, within-ESC characters will contain increasingly less phylogenetic signal as a lineage evolves, owing to a high frequency of reversals in state transitions, in much the same way that saturation of third site substitutions in DNA codons (permitted by redundancy in the genetic code) leads to high rates of homoplasy and a loss of phylogenetic information at that site. In a comparative character analysis, loss of phylogenetic signal and high homoplasy in a set of functionally related characters could signal the presence of an ESC. It would also imply that the lineage is sufficiently old to have accrued multiple state transitions, accounting for the high rate of homoplasy.

Wake (1991) has shown that homoplasy is ubiquitous in some lineages, such as the plethodontid salamanders, and that this homoplasy can be underlain by several different causal, or mechanistic, processes. Such high rates of homoplasy suggest that a high degree of character isotropy has been retained in this speciose clade, but that other clades might vary in the degree to which isotropy is possible. According to Sanderson and Donoghue (1989), the likelihood of homoplasy increases as the number of possible states of a character decreases, therefore, within-ESC character homoplasy might be further promoted owing to limitations on character state variation imposed by internal selection within the ESC complex.

Phylogenetic Signal

Within-ESC phylogenetic signal implies that character states are stable within clades. As such, sub-ESC phenotypes are recognizable within the ESC group, and these phenotypes correspond to phylogenetically delineated groups of related species (Fig. 6). This pattern might be explained in several ways:

1. *Character space isotropy and local adaptation.* It is possible that character space is isotropic and that state transitions are possible in any direction. However, some characters might be more or less sensitive to external, environmental selection pressures so that adaptation to local conditions is possible (within design limits imposed by the proper function). Thus we would expect the evolution of phenotypes optimized for local environmental conditions. As long as each clade is *uniquely* associated with a particular environment, the distribution of character states among species will provide strong phylogenetic signal. Implicit in this scenario is that differences among phenotypes reflect adaptive, functional differences discernible by functional analysis, and that clades and environments are tightly correlated (Fig. 7). However, evolutionary stability of such sub-ESC phenotypes is likely to be fleeting; temporal changes in environment and occupation of similar environments by more than one clade would promote homoplasy (reversals and convergences, respectively), mosaic character evolution, and the ultimate loss of phylogenetic signal among ESC characters (situation 1 in the previous section). Therefore, for such a pattern to prevail through evolutionary time, one must posit additional restrictions on character evolution.

2. *Local adaptation and vicariance.* Clades isolated by vicariance could evolve under uniquely different, stable environments such that locally adaptive, clade-specific phenotypes evolve. Given the potential scale of vicariance events, it is plausible that the isolated environments would remain stable over evolutionary time, at least in those attributes that affect the ESC phenotype. For example, subdivision of an ancestral clade by mountain formation could result in mesic- and xeric-adapted descendant clades owing to the creation of a "rain shadow." As long as the ESC remains functional within these new environments, local adaptation of the ESC phenotype is possible and each phenotype would be uniquely associated with a separate clade. This situation differs from (1) only in coarseness of scale. It assumes that the relevant environmental selection pressures (i.e., those to which within-ESC characters are responding) are those affected by the vicariance event. These conditions are further assumed to be stable in the long term owing to their coarse scale.

In a related situation, character evolution within the ESC might be subject to a trend imposed by cladogenesis associated with an environmental gradient (temporal or spatial). For example, expansion in the range of an ancestral taxon along a latitudinal gradient could be punctuated by occasional cladogenesis and *in situ* adaptive radiation. Temperate-adapted and tropical-adapted sub-ESC phenotypes could evolve if adaptive radiations occur at each end of the range, and these would be stable to the extent that each clade remains within its associated biome. In the case of clinal variation generally, although an upper limit is imposed by internal selection of the ESC, character variation will occur along a trajectory and homoplasy is not expected. As noted previously, sub-ESC phenotypic differences would reflect adaptive, functional differences that enhance performance of the proper function in the local, associated environment. Phenotypes would be correlated with environments because, in all cases, the sub-ESC phenotypes are maintained by external stabilizing selection.

Essential features of these scenarios are clade–environment fidelity and coarseness in scale of environmental differences to which ESC characters are adaptively sensitive (e.g., temperate vs. tropical, xeric vs. mesic). As long as these requirements are met, sub-ESC phenotypes could remain stable and phylogenetically informative, even with character state isotropy.

3. *Anisotropy of ESC character space.* Anisotropy of character space implies that not all state transitions are equally likely, or possible (e.g., character state transition $a \rightarrow a'$ can occur, but not $a' \rightarrow a$). Such anisotropy implies underlying evolutionary constraint, that is, bias or limitations of the generative systems that produce particular phenotypic states (see McKitrick, 1993; Schwenk, 1995a for the necessity of restricting constraint concepts to characters and not to organisms). Constraints could evolve at the inception of a novel character state, but perhaps are more likely to occur with reproductive isolation of a clade evincing the novel state.

This scenario accurately describes much of the pattern observed in the Case Study ESC (Fig. 2). Iguanidae and Agamidae, for example, exhibit unique sub-ESC phenotypes, but species in each taxon often occupy indistinguishable environments. There is little evidence of character state mosaicism expected with isotropy, and most phenotypic differences evident in their lingual feeding systems either have no apparent functional/performance consequence, or slight functional differences observed have no (discernible) adaptive significance in the context of present, ESC-associated environments (Schwenk and Throckmorton, 1989). For example, agamids and iguanids differ in numerous details of tongue and hyobranchium anatomy (Table I; Fig. 3; Schwenk, 1988; Smith, 1988) and slightly in the kinematics of tongue protrusion during lingual feeding (Schwenk and Throckmorton, 1989), however there is no evidence that these phenotypic differences are manifest in performance differences reflecting adaptation to local conditions. Indeed, two of the species studied by Schwenk and Throckmorton (*Uromastyx aegyptius* (Agamidae) and *Sauromalus obesus* (Iguanidae) are often regarded as Old and New World ecological counterparts—they are similar in size, occur in northern hemisphere temperate deserts, occupy similar rock outcrops, use similar, crevice-wedging defense behavior, and are unusual in being folivorous herbivores—yet each retains ESC traits characteristic of its particular clade rather than showing any convergent adaptation in ESC character states (as evident in other, non-ESC characters). Thus, the pattern suggests that most sub-ESC phenotypic differences arose, for whatever reasons, early in the origin of each clade and became fixed owing to constraints on character evolution (character space anisotropy) acting to reduce homoplasy in ESC characters (either through reversal or through convergence).

4. *The addition of new ESC characters.* Finally, in addition to creating sub-ESC phenotypes through modification of existing ESC character states, it is possible to create a novel phenotype by addition of entirely new characters. Such characters must be integrated into the ESC, which requires that the proper function be maintained, and they must also be evolutionarily stabilized in a manner consistent with the foregoing discussion. The addition of novel characters would be likely to occur within the context of other conditions discussed previously and is therefore not a mutually exclusive category for the origin of sub-ESC phenotypes. For example, in the family Chamaeleonidae there is evidence that its unique, sub-ESC phenotype (projectile tongue) arose ancestrally as an *adaptive* response to certain conditions (see Case Study). However, the chamaeleonid system incorporates both new states of preexisting ESC characters (i.e., new states of characters shared among all iguanians, e.g., modifications of the lingual process and hyoglossus muscles, etc.) *and* the addition of novel, autapomorphic characters unique to chameleons (e.g., insertion of a ballistic phase into the kinematic sequence

of tongue protrusion during lingual feeding; Bramble and Wake, 1985; Schwenk and Bell, 1988; Schwenk and Throckmorton, 1989; Smith, 1988). The addition of a ballistic phase to the ancestral feeding sequence during the evolution of chameleons would have modified internal functionality rules of the entire system and therefore, the mechanistic requirements of the projectile system would have driven a "reworking" of the ancestral phenotype through the action of internal selection. Stability of the chamaeleonid, sub-ESC phenotype (and its inherent phylogenetic signal) is, probably a combined result of historical local adaptation with subsequent character state anisotropy (probably occurring mostly at the origin of the clade) and very strong, internal selection rules unique to the group and stemming directly from the origin of novel characters integrated into the ESC.

Given that internal selection is somewhat modified in chameleons relative to other generalized (ancestral-type) iguanians, it might be argued that chameleons embody a distinct ESC that originated by "budding" from the ancestral, lingual-feeding ESC. This potentially represents an additional mode for the evolutionary origin of an ESC (ESC budding or fragmentation). An alternative, however, is to consider all iguanians within a single lingual feeding ESC because the chameleon system merely represents one extreme of a continuum in lingual prey prehension within the group (Schwenk and Bell, 1988), *all of which share a common proper function* and the fundamental phenotypic traits necessary to perform that function. Despite the projection phase, there has been no fundamental shift in the manner with which chameleons apprehend prey. On the other hand, the mutational connectivity postulate is weakened in this example, because it is possible that adaptive evolution for its proper function may not be able to turn a chameleon feeding system into a generalized iguanian feeding system. We feel that this situation requires further scrutiny to decide whether the ESC concept, as proposed here, is too narrow to capture this pattern of phenotypic evolution, or whether it is useful to recognize the chameleon system as a derivative, but distinct, ESC.

REFERENCES

Abu-ghalyun, Y. L., Greenwald, L., Hetherington, T. E., and Gaunt, A. S., 1988, The physiological basis of slow locomotion in chameleons, *J. Exp. Zool.* **245:**225–231.

Alberch, P., 1982, Developmental constraints in evolutionary processes, in: *Evolution and Development* (J. T. Bonner, Ed.), pp. 313–332, Springer Verlag, Berlin.

Allee, W. C., Park, O., Emerson, A. E., Park, T., and Schmidt, K. P., 1949, *Principles of Animal Ecology*, W. B. Saunders, Philadelphia.

Amundson, R., and Lauder, G. V., 1994, Function without purpose: The uses of causal role function in evolutionary biology, *Biol. Philos.* **9:**443–469.

Antonovics, J., and v. Tienderen, P. H., 1991, Ontoecogenophyloconstraints? The chaos of constraint terminology, *Trends in Ecol. Evol.* **6:**166–168.

Arthur, W., 1997, *The Origin of Animal Body Plans*, Cambridge Univ. Press, Cambridge.

Barel, C. D. N., 1993, Concepts of an architectonic approach to transformation morphology, *Acta Biotheor.* **41:**345–381.

Bauer, A. M., 1992, Lizards, in: *Reptiles and Amphibians* (H. G. Cogger and R. G. Zweifel, Eds.), pp. 126–173, Smithmark, New York.

Bell, D. A., 1990, Kinematics of prey capture in the chameleon, *Zool. Jb. Physiol.* **94:**247–260.

Bels, V. L., and Goose, V., 1989, A first report of relative movements within the hyoid apparatus during feeding in *Anolis equestris* (Reptilia: Iguanidae), *Experientia* **45:**1088–1091.

Bels, V. L., Chardon, M., and Kardong, K. V., 1994, Biomechanics of the hyolingual system in Squamata, in: *Biomechanics of Feeding in Vertebrates. Advances in Comparative & Environmental Physiology* (V. L. Bels, M. Chardon and P. Vandewalle, Eds.), pp. 197–240, Springer Verlag, Berlin.

Billo, R., and Wake, M. H., 1987, Tentacle development in *Dermophis mexicanus* (Amphibia, Gymnophiona) with an hypothesis of tentacle origin, *J. Morph.* **192**:101–111.

Bock, W. J., 1963, Evolution and phylogeny in morphologically uniform groups, *Amer. Nat.* **97**:265–285.

Bock, W. J., 1964, Kinetics of the avian skull, *J. Morph.* **114**:1–41.

Bock, W. J., and von Wahlert, G., 1965, Adaptation and the form–function complex, *Evolution* **19**:269–299.

Bramble, D. M., and Wake, D. B., 1985, Feeding mechanisms of lower tetrapods, in: *Functional Vertebrate Morphology* (M. Hildebrand, D. M. Bramble, K. F. Liem and D. B. Wake, Eds.), pp. 230–261, Harvard University Press, Cambridge, MA.

Burghardt, G. M., 1970, Chemical perception in reptiles, in: *Communication by Chemical Signals* (J. W. Johnston, D. G. Moulton and A. Turk, Eds.), pp. 241–308, Appleton-Century-Crofts, New York.

Carroll, R. L., 1988, *Vertebrate Paleontology and Evolution*, W. H. Freeman, New York.

Cooper, W. E., 1996, Preliminary reconstructions of nasal chemosensory evolution in Squamata. *Amph.-Rept.* **17**:395–415.

Crandall, K. A., and Hillis, D. M., 1997, Rhodopsin evolution in the dark, *Nature* **387**:667–668.

Cundall, D., 1995, Feeding behaviour in *Cylindrophis* and its bearing on the evolution of alethinophidian snakes, *J. Zool. London* **237**:353–376.

Darwin, C., 1859, *On the Origin of Species by Means of Natural Selection or the Preservation of Favored Races in the Struggle for Life*, Murray, London.

deBraga, M., and Rieppel, O., 1997, Reptile phylogeny and the interrelationships of turtles, *Zool. J. Linn. Soc.* **120**:281–354.

Delheusy, V., and Bels, V. L., 1992, Kinematics of feeding behavior in *Oplurus cuvieri* (Reptilia: Iguanidae), *J. Exp. Biol.* **179**:155–186.

Delheusy, V., Toubeau, G., and Bels, V. L., 1994, Tongue structure and function in *Oplurus cuvieri* (Reptilia: Iguanidae), *Anat. Rec.* **238**:263–276.

Díaz-Uriarte, R., and Garland, J. T., 1996, Testing hypotheses of correlated evolution using phylogenetically independent contrasts: Sensitivity to deviations from Brownian motion, *Syst. Biol.* **45**:27–47.

Dullemeijer, P., 1956, The functional morphology of the head of the common viper, *Vipera berus* (L.), *Arch. Neerl. Zool.* **11**:386–497.

Dullemeijer, P., 1959, A comparative functional-anatomical study of the heads of some Viperidae, *Morph. Jb.* **99**:881–985.

Dullemeijer, P., 1974, *Concepts and Approaches in Animal Morphology*, Van Gorcum, Assen, The Netherlands.

Dullemeijer, P., 1980, Functional morphology and evolutionary biology, *Acta Biotheoretica* **29**:151–250.

Dullemeijer, P., 1989, On the concept of integration in animal morphology, in: *Trends in Vertebrate Morphology* (H. Splechtna and H. Hilgers, Eds.), pp. 3–18, Gustav Fischer Verlag, Stuttgart.

Emerson, A. E., 1949, Adaptation, in: *Principles of Animal Ecology* (by W. C. Allee, O. Park, A. E. Emerson, T. Park and K. P. Schmidt), pp. 630–690, W. B. Saunders, Philadelphia.

Estes, R., 1983, The fossil record and early distribution of lizards, in: *Advances in Herpetology and Evolutionary Biology* (A. G. J. Rhodin and K. Miyata, Eds.), pp. 365–398, Museum of Comparative Zoology, Cambridge, MA.

Estes, R., de Queiroz, K., and Gauthier, J., 1988, Phylogenetic relationships within Squamata, in: *Phylogenetic Relationships of the Lizard Families* (R. Estes and G. Pregill, Eds.), pp. 119–281, Stanford University Press, Stanford, CA.

Felsenstein, J., 1985, Phylogenies and the comparative method, *Amer. Nat.* **125**:1–15.

Frazzetta, T. H., 1986, The origin of amphikinesis in lizards, *Evol. Biol.* **20**:419–461.

Galis, F., 1992, A model for biting in the pharyngeal jaws of a cichlid fish: *Haplochromis piceatus*, *J. Theor. Biol.* **155**:343–368.

Galis, F., 1996, The application of functional morphology to evolutionary studies, *Tr. Ecol. Evol.* **11**:124–129.

Galis, F., personal communication.

Galis, F., and Drucker, E. G., 1996, Pharyngeal biting mechanics in centrarchid and cichlid fishes: Insights into a key evolutionary innovation, *J. Evol. Biol.* **9**:641–670.

Gans, C., 1969a, Comments on inertial feeding, *Copeia* **1969**:855–857.

Gans, C., 1969b, Functional components versus mechanical units in descriptive morphology, *J. Morph.* **128**:365–368.

Garland, T., Jr., 1992, Rate tests for phenotypic evolution using phylogenetically independent contrasts, *Am. Nat.* **140**:509–519.

Garland, T., Jr., Harvey, P. H., and Ives, A. R., 1992, Procedures for the analysis of comparative data using phylogenetically independent contrasts, *Syst. Biol.* **41**:18–32.

Gatesy, S. M., and Dial, K. P., 1996, Locomotor modules and the evolution of avian flight, *Evolution* **50**:331–340.

Gauthier, J., Estes, R., and de Queiroz, K., 1988, A phylogenetic analysis of Lepidosauromorpha, in: *Phylogenetic Relationships of the Lizard Families* (R. Estes and G. Pregill, Eds.), pp. 15–98, Stanford University Press, Stanford, CA.

Gerhart, J., and Kirschner, M., 1997, *Cells, Embryos and Evolution*, Blackwell Science, Malden, MA.

Goodrich, E. S., 1906, Notes on the development, structure and origin of the median and paired fins of fish, *Quart. J. Microsc. Sci.* **50**:333–376.

Goodrich, E. S., 1913, Metameric segmentation and homology, *Quart. J. Microsc. Sci.* **59**:227.

Goose, V., and Bels, V. L., 1992, Kinematic and functional analysis of feeding behavior in *Lacerta viridis* (Reptilia: Lacertidae), *Zool. Jb. Anat.* **122**:187–202.

Gorniak, G. C., Rosenberg, H. I., and Gans, C., 1982, Mastication in the tuatara *Sphenodon punctatus* (Reptilia: Rhynchocephalia): Structure and activity of the motor system, *J. Morph.* **171**:321–353.

Gould, S. J., and Eldredge, N., 1977, Punctuated equilibria: The tempo and mode of evolution reconsidered, *Paleobiology* **3**:115–151.

Graham, J. B., 1997, *Air Breathing Fishes: Evolution, Diversity, and Adaptation*, Academic Press, San Diego, CA.

Greene, H. W., 1982, Dietary and phenotypic diversity in lizards: Why are some organisms specialized? in: *Environmental Adaptation and Evolution* (D. Mossakowski and G. Roth, Eds.), pp. 107–128, Gustav Fischer, Stuttgart.

Greer, A. E., 1989, *The Biology and Evolution of Australian Lizards*, Surrey Beatty and Sons, Chipping Norton, Australia.

Hall, B. K., 1992, *Evolutionary Developmental Biology*, Chapman and Hall, London.

Hall, B. K., 1996, Baupläne, phylotypic stages and constraint. Why there are so few types of animals, in: *Evolutionary Biology*, Vol. 29 (M. K. Hecht, R. J. MacIntyre and M. T. Clegg, Eds.), pp. 215–261, Plenum Press, New York.

Halpern, M., 1992, Nasal chemical senses in reptiles: Structure and function, in: *Biology of the Reptilia*, Vol. 18 (C. Gans and D. Crews, Eds.), pp. 423–523, University of Chicago Press, Chicago.

Hansen, T. F., and Martins, E. P., 1996, Translating between microevolutionary process and macroevolutionary patterns: The correlation structure of interspecific data, *Evolution* **50:**1404–1417.

Harvey, P. H., and Pagel, A. D., 1991, *The Comparative Method in Evolutionary Biology*, Oxford University Press, Oxford.

Hennig, W., 1966, *Phylogenetic Systematics*, University of Illinois Press, Urbana.

Herrel, A., Cleuren, J., and DeVree, F., 1995, Prey capture in the lizard *Agama stellio*, *J. Morph.* **224:**313–329.

Hiiemae, K. M., and Crompton, A. W., 1985, Mastication, food transport, and swallowing, in: *Functional Vertebrate Morphology* (M. Hildebrand, D. M. Bramble, K. F. Liem, and D. B. Wake, Eds.), pp. 262–290, Harvard University Press, Cambridge, MA.

Holsinger, K., personal communication.

Hughes, A. L., 1994, The evolution of functionally novel proteins after gene duplication, *Proc. R. Soc. Lond. B* **256:**119–124.

Jaksic, F. M., and Schwenk, K., 1983, Natural history observations on *Liolaemus magellanicus*, the southernmost lizard in the world, *Herpetologica* **39:**457–461.

Jaksic, F. M., Fuentes, E. R., and Yanez, J. L., 1979, Two types of adaptation of vertebrate predators to their prey, *Arch. Biol. Med. Exper.* **12:**143–152.

Jones, T., 1995, *Evolutionary Algorithms, Fitness Landscapes and Search*, PhD Dissertation, University of New Mexico., Albuquerque, NM.

Kimura, M., 1983, *The Neutral Theory of Molecular Evolution*, Cambridge University Press, Cambridge, UK.

Kley, N. J., and Brainerd, E. L., 1996, Internal concertina swallowing: A critical component of alethinophidian feeding systems, *Amer. Zool.* **36:**81A.

Klingenberg, C. P., and Nijhout, H. F., 1998, Competition among growing organs and developmental control of morphological asymmetry, *Proc. R. Soc. Lond. B* **265:**1135–1139.

Kraklau, D. M., 1991, Kinematics of prey capture and chewing in the lizard *Agama agama* (Squamata: Agamidae), *J. Morph.* **210:**195–212.

Lauder, G. V., and Liem, K. F., 1989, The role of historical factors in the evolution of complex organismal functions, in: *Complex Organismal Functions: Integration and Evolution in Vertebrates* (D. B. Wake and G. Roth, Eds.), pp. 63–78, John Wiley and Sons, New York.

Lee, M. S. Y., 1997, Reptile relationships turn turtles, *Nature* **389:**245–246.

Liem, K. F., 1988, Form and function of lungs: The evolution of air breathing mechanisms, *Amer. Zool.* **28:**739–759.

Losos, J. B., and Greene, H. W., 1988, Ecological and evolutionary implications of diet in monitor lizards, *Biol. J. Linn. Soc.* **35:**379–407.

Macey, J. R., Larson, A., Ananheva, N. B., and Papenfuss, T. J., 1997, Evolutionary shifts in three major structural features of the mitochondrial genome among iguanian lizards, *J. Mol. Evol.* **44:**660–674.

Maddison, W. P., and Maddison, D. R., 1992, *MacClade version 3. Analysis of Phylogeny and Character Evolution*, Sinauer Associates, Sunderland, MA.

Martins, E. P., 1996a, Conducting phylogenetic comparative studies when phylogeny is not known, *Evolution* **50:**12–22.

Martins, E. P., 1996b, *Phylogenies and the Comparative Method in Animal Behavior*, Oxford University Press, Oxford.

Martins, E. P., and Hansen, T. F., 1997, Phylogenies and the comparative method: A general approach to incorporating phylogenetic information into the analysis of interspecific data, *Amer. Nat.* **149:**646–667.

Mason, R. T., 1992, Reptilian pheromones, in: *Hormones, Brain and Behavior. Biology of the Reptilia*, Vol. 18 (C. Gans and D. Crews, Eds.), pp. 114–228, University of Chicago Press, Chicago.

Maynard-Smith, J., Burian, R., Kauffman, S., Alberch, P., Campbell, J., Goodwin, B., Lande, R., Raup, D., and Wolpert, L., 1985, Developmental constraints and evolution, *Quart. Rev. Biol.* **60:**265–287.

McKitrick, M. C., 1993, Phylogenetic constraint in evolutionary theory: Has it any explanatory power? *Anu. Rev. Ecol. Syst.* **24:**307–330.

Millikan, R. G., 1984, *Language, Thought, and Other Biological Categories*. MIT Press, Cambridge, MA.

Minelli, A., and Peruffo, B., 1991, Developmental pathways, homology and homonomy in metameric animals, *J. Evol. Biol.* **4:**429–445.

Müller, G. B., and Wagner, G. P., 1991, Novelty in evolution: Restructuring the concept, *Annu. Rev. Ecol. Syst.* **22:**229–256.

Nijhout, H. F., and Emlen, D. J., 1998, Competition among parts in the development and evolution of insect morphology, *Proc. Natl. Acad. Sci. USA* **95:**3685–3689.

Nowak, M. A., Boerlijst, M. C., Cooke, J., and Maynard Smith, J., 1997, Evolution of genetic redundancy, *Nature* **388:**167–171.

Ohno, S., 1970, *Evolution by Gene Duplication*, Springer-Verlag, New York.

O'Reilly, J. C., Nussbaum, R. A., and Boone, D., 1996, Vertebrate with protrusible eyes, *Nature* **382:**33.

Pianka, E. R., 1986, *Ecology and Natural History of Desert Lizards*, Princeton University Press, Princeton, New Jersey.

Pigliucci, M., Schlichting, C. D., Jones, C. S., and Schwenk, K., 1996, Developmental reaction norms: The interactions among allometry, ontogeny and plasticity, *Plant Species Biol.* **11:**69–85.

Pough, F. H., 1973, Lizard energetics and diet, *Ecology* **54:**837–844.

Riedl, R., 1978, *Order in Living Organisms: A Systems Analysis of Evolution*, John Wiley & Sons, New York.

Rieppel, O., and deBraga, M., 1997, Turtles as diapsid reptiles, *Nature* **384:**453–455.

Rollo, C. D., 1995, *Phenotypes*, Chapman and Hall, London.

Rose, M. R., and Lauder, G. V., 1996, *Adaptation*, Academic Press, San Diego.

Roth, G., and Wake, D. B., 1985, Trends in the functional morphology and sensorimotor control of feeding behavior in salamanders: An example of the role of internal dynamics in evolution, *Acta Biotheoretica* **34:**175–192.

Roth, G., and Wake, D. B., 1989, Conservatism and innovation in the evolution of feeding in vertebrates, in: *Complex Organismal Functions: Integration and Evolution in Vertebrates* (D. B. Wake and G. Roth, Eds.), pp. 7–21, John Wiley & Sons, New York.

Roux, W., 1881, *Kampf der Theile im Organismus*, Jena, Leipzig.

Rowe, T., 1996, Coevolution of the mammalian middle ear and neocortex, *Science* **273:**651–654.

Sanderson, M. J., and Donoghue, M. J., 1989, Patterns of variation in levels of homoplasy, *Evolution* **43:**1781–1795.

Schaefer, S. A., and Lauder, G. V., 1996, Testing historical hypotheses of morphological change: Biomechanical decoupling in loricarioid catfishes, *Evolution* **50:**1661–1675.

Schlichting, C. D., and Pigliucci, M., 1998, *Phenotypic Evolution: A Reaction Norm Perspective*, Sinauer Assoc., Sunderland, MA.

Schmalhausen (Shmal'gauzen), I. I., 1949, *Factors of Evolution. The Theory of Stabilizing Selection*, Blakiston Co., Philadelphia.

Schwenk, K., 1986, Morphology of the tongue in the tuatara, *Sphenodon punctatus* (Reptilia: Lepidosauria), with comments on function and phylogeny, *J. Morph.* **188:**129–156.

Schwenk, K., 1988, Comparative morphology of the lepidosaur tongue and its relevance to squamate phylogeny, in: *Phylogenetic Relationships of the Lizard Families* (R. Estes and G. Pregill, Eds.), pp. 569–598, Stanford University Press, Stanford, CA.

Schwenk, K., 1993, The evolution of chemoreception in squamate reptiles: A phylogenetic approach, *Brain Behav. Evol.* **41:**124–137.

Schwenk, K., 1994a, Comparative biology and the importance of cladistic classification: A case study from the sensory biology of squamate reptiles, *Biol. J. Linn. Soc.* **52:**69–82.

Schwenk, K., 1994b, Systematics and subjectivity: The phylogeny and classification of iguanian lizards reconsidered, *Herp. Rev.* **25:**53–57.

Schwenk, K., 1994c, Why snakes have forked tongues, *Science* **263:**1573–1577.

Schwenk, K., 1995b, Of tongues and noses: Chemoreception in lizards and snakes, *Trends Ecol. Evol.* **10:**7–12.

Schwenk, K., 1995a, A utilitarian approach to evolutionary constraint, *Zoology* **98:**251–262.

Schwenk, K., 1996, Why snakes flick their tongues, *Amer. Zool.* **36:**84A.

Schwenk, K., in press, Feeding in lepidosaurs, in: *Feeding in Tetrapod Vertebrates: Form, Function, Phylogeny* (K. Schwenk, Ed.), Academic Press, San Diego.

Schwenk, K., and Bell, D. A., 1988, A cryptic intermediate in the evolution of chameleon tongue projection, *Experientia* **44:**697–700.

Schwenk, K., and Throckmorton, G. S., 1989, Functional and evolutionary morphology of lingual feeding in squamate reptiles: Phylogenetics and kinematics, *J. Zool., Lond.* **219:**153–175.

Schwenk, K., and Wanger, G. P., in preparation, The reconciliation of selection and constraint.

Simpson, G. G. 1953, *The Major Features of Evolution*, Columbia Univ. Press, New York.

Sinnott, E. W., 1946, Substance or system: The riddle of morphogenesis, *Amer. Nat.* **80:** 497–505.

Smith, K. K., 1984, The use of the tongue and hyoid apparatus during feeding in lizards (*Ctenosaura similis* and *Tupinambis nigropunctatus*), *J. Zool., Lond.* **202:**115–143.

Smith, K. K., 1986, Morphology and function of the tongue and hyoid apparatus in *Varanus* (Varanidae, Lacertilia), *J. Morph.* **187:**261–287.

Smith, K. K., 1988, Form and function of the tongue in agamid lizards with comments on its phylogenetic significance, *J. Morph.* **196:**157–171.

Smith, K. K., and Mackay, K. A., 1990, The morphology of the intrinsic tongue musculature in snakes (Reptilia, Ophidia): Functional and phylogenetic implications, *J. Morph.* **205:**307–324.

Stearns, S. C., 1993, The evolutionary links between fixed and variable traits, *Acta Palaeon. Pol.* **38:**1–17.

Throckmorton, G. S., 1980, The chewing cycle in the herbivourous lizard *Uromastix aegyptius* (Agamidae), *Archs. Oral Biol.* **25:**225–233.

Urbani, J.-M., and Bels, V. L., 1995, Feeding behaviour in two scleroglossan lizards: *Lacerta viridis* (Lacertidae) and *Zonosaurus laticaudatus* (Cordylidae), *J. Zool., Lond.* **236:** 265–290.

van der Klaauw, C. J., 1945, Cerebral skull and facial skull. A contribution to the knowledge of skull structure, *Arch. Neerl. Zool.* **7:**16–37.

Vermeij, G., 1974, Adaptation, versatility and evolution, *Syst. Zool.* **22:**466–477.

von Bertalanffy, L., 1952, *Problems of Life. An Evaluation of Modern Biological Thought*, Watts, London.

Waddington, C. H., 1957, *The Strategy of the Genes*, MacMillan Co., New York.

Wagner, G. P., 1986, The systems approach: An interface between development and population genetic aspects of evolution, in: *Patterns and Processes in the History of Life* (D. M. Raup and D. Jablonski, Eds.), pp. 149–165, Springer-Verlag, Berlin.

Wagner, G. P., 1989, The origin of morphological characters and the biological basis of homology, *Evolution* **43:**1157–1171.

Wagner, G. P., 1994, Homology and the mechanisms of development, in: *Homology: The Hierarchical Basis of Comparative Biology* (B. K. Hall, Ed.), pp. 273–299, Academic Press, San Diego.

Wagner, G. P., in press, *The Character Concept in Evolutionary Biology*, Academic Press, San Diego.

Wagner, G. P., and Altenberg, L., 1996, Complex adaptations and the evolution of evolvability, *Evolution* **50:**967–976.

Wagner, G. P., Booth, G., and Bagheri-Chaichian, H., 1997, A population genetic theory of canalization, *Evolution* **51:**329–347.

Wake, D. B., 1991, Homoplasy: The result of natural selection, or evidence of design limitations? *Amer. Nat.* **138:**543–567.

Wake, M. H., 1985, The comparative morphology and evolution of the eyes of caecilians (Amphibia, Gymnophiona), *Zoomorphology* **105:**277–295.

Wake, M. H., 1992, "Regressive" evolution of special sensory organs in caecilians (Amphibia: Gymnophiona): Opportunity for morphological innovation, *Zool. Jb. Anat.* **122:**325–329.

Wake, D. B., and Larson, A., 1987, Multidimensional analysis of an evolving lineage, *Science* **238:**42–48.

Wake, D. B., Roth, G., and Wake, M. H., 1983, On the problem of stasis in organismal evolution, *J. Theor. Biol.* **101:**21–224.

Walker, W. F., Jr., and Liem, K. F., 1994, *Functional Anatomy of the Vertebrates*, Saunders, Philadelphia.

Walls, G. L., 1942, *The Vertebrate Eye and its Adaptive Radiation*, Cranbrook Institute of Science, Bloomfield Hills, MI.

Weismann, A., 1896, *On Germinal Selection as a Source of Definite Variation*, 2nd ed. (English translation, Religious and Science Library, London, 1902).

Weiss, K. M., 1990, Duplication with variation: Metameric logic in evolution from genes to morphology, *Yearb. Phys. Anthropol.* **33:**1–23.

Whyte, L. L., 1965, *Internal Factors in Evolution*, George Braziller, New York.

Wimsatt, W. C., and Schank, J. C., 1988, Two constraints on the evolution of complex adaptations and the means for their avoidance, in: *Evolutionary Progress* (M. Nitecki and K. Nitecki, Eds.), pp. 231–273, Univ. of Chicago Press, Chicago.

Index